THE CLIMATE CHANGE CHALLENGE AND THE FAILURE OF DEMOCRACY

Recent Titles in Politics and the Environment

Gaia's Revenge: Climate Change and Humanity's Loss
P. H. Liotta and Allan W. Shearer

Ignoring the Apocalypse: Why Planning to Prevent Environmental
Catastrophe Goes Astray
David Howard Davis

THE CLIMATE CHANGE CHALLENGE AND THE FAILURE OF — DEMOCRACY —

DAVID SHEARMAN AND

JOSEPH WAYNE SMITH

Politics and the Environment
P. H. Liotta, Series Editor

Westport, Connecticut
London

Library of Congress Cataloging-in-Publication Data

Shearman, David.

The climate change challenge and the failure of democracy / David Shearman and Joseph Wayne Smith.

       p. cm. — (Politics and the environment, ISSN 1932–3484)

   Includes bibliographical references and index.

   ISBN-13: 978–0–313–34504–3 (alk. paper)

   ISBN-10: 0–313–34504–X (alk. paper)

  1.   Democracy. 2.   Authoritarianism. 3.   Climatic changes—Political aspects. 4.   Environmental policy.   I. Smith, Joseph Wayne. II. Title.

   JC423.S496 2007

   363.738'74561—dc22   2007018275

British Library Cataloguing in Publication Data is available.

Library of Congress Catalog Card Number: 2007018275

ISBN-13: 978–0–313–34504–3
ISBN-10: 0–313–34504–X
ISSN: 1932–3484

First published in 2007

Praeger Publishers, 88 Post Road West, Westport, CT 06881
An imprint of Greenwood Publishing Group, Inc.
www.praeger.com

Printed in the United States of America

The paper used in this book complies with the Permanent Paper Standard issued by the National Information Standards Organization (Z39.48–1984).

10 9 8 7 6 5 4 3 2 1

*This book is dedicated to all who work for a truly equitable
and environmentally sustainable world.*

Contents

Series Foreword

The key focus of the Praeger Politics and Environment series is to explore the interstices between environment, political, and security impacts in the twenty-first century. To those intimately involved with these issues, their immediacy and importance are obvious. What is not obvious to many, nonetheless—including those involved in making decisions that affect our collective future—is how these three critical issues are in constant conflict and frequently clash. Today, more so than at any other time in human history, intersecting environment, political, and security issues profoundly impact our lives and the lives of those who are to come.

In examining the complex interdependence of these three impact effects, the study of environmental and security issues should recognize several distinct and pragmatic truths. One, international organizations today are established for and focus best on security issues. Thus, while it remains difficult to address environmental threats, challenges, and vulnerabilities for these organizations, it makes imminently better sense to reform what we have rather than constantly invent the "new" organization that may be no better equipped to handle current and future challenges. Secondly, the introduction of new protocols must continue to be created, worked into signature, and managed under the leadership of states through international organizations and cooperative regimes. Finally, and incorporating the reality of these previous two truths, we should honestly recognize that environmental challenges can best be presented in terms that relate to security issues. To that end, it is sensible to depict environmental challenges in language that is understandable to decision makers most familiar with security impacts and issues.

There is benefit and danger in this approach, of course. Not all security issues involve direct threats; some security issues, as with some political processes, are far more nuanced, more subtle, and less clearly evident. I would argue further—as I have been arguing for several decades now—that it remains a tragic mistake to couch all security issues in terms of threat. To the contrary, what I term "creeping vulnerabilities"—climate change, population growth, disease, climate change, scarcity of water and other natural resources, decline in food production, access, and availability, soil erosion and desertification, urbanization and pollution, and the lack of effective warning systems—can come to have a far more devastating impact effect if such issues are ignored and left unchecked over time. In the worst possible outcomes, vulnerabilities left unchecked over time *will* manifest themselves as threats.

In its most direct, effective, and encompassing assessment, environmental security centers on a focus that seeks the best response to changing environmental conditions that have the potential to reduce stability and affect peaceful relationships, and—if left unchecked—could lead to the outbreak of conflict. This working definition, therefore, represents the vital core of the Praeger Politics and the Environment series.

Environmental security emphasizes the sustained viability of the ecosystem, while recognizing that the ecosystem itself is perhaps the ultimate weapon of mass destruction. In 1556 in Shensi province, for example, tectonic plates shifted and by the time they settled back into place, 800,000 Chinese were dead. Roughly 73,500 years ago, a volcanic eruption in what is today Sumatra was so violent that ash circled the earth for several years, photosynthesis essentially stopped, and the precursors to what is today the human race amounted to only several thousand survivors worldwide. The earth itself, there can be little doubt, is the ultimate weapon of mass destruction. Yet from an alternate point of view, mankind itself is the ultimate threat to the earth and the earth's ecosystem.

Three decades ago, the environmentalist Norman Myers wrote that national security is about far more than fighting forces and weaponry. National security must also include issues of environment and environmental impact—from watersheds to climate impact—and these factors must figure in the minds of military exerts and political leaders. Myers' words today remain as prophetic, and deadly accurate, as ever.

In this latest volume of the Politics and Environment series, professor emeritus of medicine David Shearman and philosopher and ecologist Joseph

Wayne Smith show a complete willingness to challenge organizational iden- tities. They forcefully argue that our system of overall political, economic, and social governance is an obstacle to confronting effectively the looming environmental crises that global climate change poses.

Shearman and Smith challenge us to reexamine how states, corporations, and consumers are driving us, literally, to the brink of disaster. In engaging considerations of the limits to growth, the separation of corporatism and governance, financial reform, legal reform, and the reclaiming of the "com- mons" for human society, they ask us to consider what is often considered unthinkable in our cosmopolitan, ideologically centered mindset. In short, Shearman and Smith argue that liberal democracy—considered sacrosanct in modern societies—is an impediment to finding ecologically sustainable solutions for the planet.

Many, of course, will find this argument untenable. But I would urge readers to listen closely to Shearman and Smith's entertaining and always thoughtful arguments. In an era of ever widening, ever deepening global- ization, liberal democracies have proven unable, or unwilling, to check the explosive growth of corporatism's power, influence, and reach. In terms that mirror Marx's thoughts on the aggregation of capital into the hands of the fewer and fewer, Shearman and Smith are nonetheless, not neo-Marxists in their argument. In some ways, their notions that liberal democracy must give way to "a form of authoritarian government by experts" reminds us as well of Plato's *Republic*.

Environmentalists often predict an Apocalypse is coming: The earth will heat up like a greenhouse. We will run out of energy. Overpopulation will lead to starvation and war. Nuclear winter will devastate all organic life. We have, of course, grown desensitized to many such prophecies of doom. Although some may find Shearman and Smith's conclusions utopian, their warnings bear close attention. As they methodically and pragmatically re- mind us throughout this book, the time for strategic reconsideration of how we run our lives—and how our lives are being run—is more pressing than ever.

P. H. Liotta
Executive Director
Pell Center for International Relations and Public Policy
Newport, Rhode Island

Preface

This book documents the near certainty of climate change, its serious consequences, and the failure of democratic societies to respond adequately. A new planet is about to be created, one that is inhospitable, producing less food and water and without the necessary ecological services to support the world's population. In February of 2007, the first part of the 4th "Report of the Intergovernmental Panel on Climate Change" was published. A consensus of 2,300 climate scientists, many of them American, reported more severe changes than in the previous report in 2001 and stressed an urgency to act.

We have known about these impending problems for several decades. Each year the certainty of the science has increased, yet we have failed to act appropriately to the threat. We have analyzed the reasons for this indolence. This understanding will lead you to ask yourself if Western civilization can survive in its present state of prosperity, health, and well-being, or will it soon suffer the fate of all previous civilizations—to become a mere page in history?

We will demand from you the reader, far more than your comprehension of the consequences of climate change and the workings of democracy. You will need to examine the limits of your introspection and the motivation bestowed upon you by biology and culture. The questions to be asked are difficult. You have a commitment to your children, but are you committed to the well-being of future generations and those you may never see, such as your great-grandchildren? If so are you prepared to change your lifestyle now? Are you prepared to see society and its governance change if this is a necessary solution?

It is salutatory to think where this issue sits in your priority list. Examine how much of your time each day is spent considering matters of importance to you. If we exclude time spent in work and sleep, what proportion of your thinking time is devoted to your career, reputation, colleagues, finances, car, future possessions, prowess, and—not least—sex, desire, and food? Of course you love your partner and children, but how much time does your brain spend on them? If your gender is female, chances are that your priorities have a different emphasis with much more time thinking of relationships and family. We may make ourselves comfortable by saying the future of humanity is a concern to us, but estimate truthfully to yourself how many seconds per day you think about it. How much time compared to your indulgencies of entertainment, television, and the delights and addictions of consumerism?

Human nature being as it is, we do not consider that these world issues threaten us until they impinge directly upon us. The crushing drought in Australia has coincided with a surge of interest in climate change. Hurricane Katrina had a similar though smaller impact on climate change discussion in the United States, but has the issue permeated the people or press of New York or the citizens of Delhi or Toronto? Has it superseded the Grammies or Oscars in public interest? The increase in wild fires in California, British Colombia, Australia, and Iberia has concentrated the minds of inhabitants because the threat is recurrent and has increased visibly. However when it happens to someone else, even though we have played a part in its genesis, it scarcely impinges on our thoughts, unlike those poor souls already subject to the inundation of a South Sea island due to rising sealevels, or the melting of the Inuit land and loss of their livelihood.

In simple terms we have a conflict of interest between our personal needs and desires and the recognition of what we must do personally to alleviate this threat. We are like many patients informed of a diagnosis of cancer. We understand the diagnosis but engage in denial. In the same way death is dismissed when it crosses our minds. Admittedly, denial has been overcome in some countries with the development of alternative fuels, but even those nations with good intent have failed to stem the rising flow of greenhouse emissions.

If conflict of interest presents a problem to all of us, this conflict is an even greater problem to those in government. This conflict explains why government rhetoric is followed by catatonia. Not only does the politician have to contend with the personal conflicts of interest that we all experience, but he or she has a career conflict over reelection, the consuming motive of most. Reelection depends upon economic growth and a booming economy, the root cause of climate change and the rapid depletion of resources available to us. The fundamental political dilemma is illustrated by the view of British Prime Minister Blair, expressed at the Economic Forum in Davos in February 2005 to the effect that if we were to put forward a solution to climate change, something that would involve drastic cuts in economic growth or standards of living, it would not matter how justified it was, it would simply not be agreed to.[1] In other words, democracy itself has a big problem. This negative response is seen to be much more vigorous when we consider the words and actions of U.S. President George Bush or Australian Prime Minister John Howard. Nothing can be done that will harm industry or jobs. Comprehension does not extend to the possibility that there may be fewer future jobs unless action is taken. In this book we provide an analysis of this situation.

This political attitude also explains the obsession with technological development. It offers a fix without having to make difficult and unpopular decisions. Climate change, like all other problems that humanity has encountered, will be solved by technology—pump the carbon dioxide underground and launch space ships with mirrors to reflect sunlight. This solution fits the paradigm of development and progress and more importantly absolves us from any sacrifice of our profligacy. But it won't work this time because there are so many interlocking problems that cannot respond to a technological fix. These problems depend upon population expansion and consumption of natural resources.

Perhaps the most important conflict of interest occurs within the corporate empires, the boiler rooms of our productive society. You will learn from our analysis that profits and responsibility to shareholders rank above all other responsibilities whatever the public face of corporate responsibility. To date, evidence suggests that the gulf cannot be bridged.

Ultimately we are enclosed in an autonomous market economy; no one can flourish outside it and the consequences of its never-ending growth are obvious to all who are prepared to think about it. This intricate form of human organization has become like the ant hill, where the brain ruling the paradigm is the collective neural tissue of all ants, working in concert and eating the world. Fortunately there are predators for ants. The autonomous human brain is "the market."

In the early chapters we introduce the reader to the scientific evidence of overwhelming environmental damage. We are now living on capital as well as interest. Climate change is not the only symptom of the stress we are putting on the earth. In fact there are many interdependent factors causing symptoms, for example the loss of productive land, decreasing fresh water, and loss of food and biological resources. Our analysis indicates common threads in the functioning of democracy in all these problems. We then define the principles of democracy in chapter 5 and explain the concept of "the commons" and the consequences of democracy failing to recognize its importance. Chapters 6 and 7 demonstrate that the inherent failures of democracy that have lead to the environmental crisis also operate in many other spheres of society. They are inherent to the operation of democracy. Furthermore, we come to share Plato's conclusion that democracy is inherently contradictory and leads naturally to authoritarianism.

In chapters 8 and 9 we argue that authoritarianism is the natural state of humanity, and it may be better to choose our elites rather than have them imposed. Indeed Plato, on seeing the sequelae of democracy's birth, observed that it is better that the just and wise should rule unwillingly, rather than those who actually want power should have it. We analyze authoritarian structures and their operation ranging from the medical intensive care unit and the Roman Catholic Church to corporatism with the conclusion that the crisis is best countered by developing authoritarian government using some of the fabric of these existing structures. The education and values of the new "elite warrior leadership" who will battle for the future of the earth is described. In chapter 10 we provide some solutions to the illnesses of democracy, and we leave you, the reader, to decide how humanity should proceed.

A PERSONAL MESSAGE FROM DAVID SHEARMAN

Every author will use an analogy to describe the hundreds of tortured hours spent writing each book. It is the excision of a painful tooth, for pain will continue until the surgery, or book, is accomplished. As a physician, scientist, and academic, I have had a happy, fulfilling life. Easing the sufferings of patients shows to me the courage, fortitude, and warmth of the individual that contrasts with the selfish abandon that humanity can display collectively. My needs are satisfied by family relationships and the fascinations of the natural world. From my daily interaction with patients, I encounter what the research studies tell us—that above a very modest income there is no relationship between happiness and income. There is no more than transient gratification in a consumerism driven by envy and wealth accumulation. These issues are well documented in our text. So my message is that all of us in wealthy democratic societies consuming resources that would require more than three planet earths to assuage, could live a simpler, fulfilling life and thereby allow our offspring to live in a sustainable world.

I have been part of the generations that created this selfish mess, and I need to spend part of my time in restitution. For all of us knowledge of this crisis demands an ethical response.

Finally, why is this book written by Australians for publication in the United States? Indeed some readers may feel that the United States is criticized unfairly. If this is so, please remember that the comments are those that have to be said to a friend. Remember also that your constitution and your publishers support freedom of speech, for which we are grateful.

Part of my medical practice was at a wonderful medical center, Yale University Medical School. I was infected by the quest for intellectual adventure and service, and by the vitality and enterprise. America epitomized the future. Today, my feeling for America, though not its citizens, is one of loss and grief. After half a century of death and brutality, the U.S. inherited untold power to shape a better world, and, when the Berlin Wall collapsed, it became a sole responsibility. Each act of human history, each civilization is a page with the same mistakes, disasters, and conclusions born of human foible and greed. History teaches us to expect no more or no less. The dream was that this empire would be different from those before. American leadership had to succeed in changing a world bent on certain ecological and possible nuclear destruction. It has failed to put self-interest aside.

The emissions of carbon dioxide in the past six years cannot be retrieved; they will continue to harm the world for many decades. As a result in the eyes of many who looked to the United States for leadership, there is bitter disappointment. The U.S. democracy that offered freedom with diminishing collective responsibility is not a model that can sustain the world. It brings recognition that democracy must be reformed. This is the motivation for this book. And the United States is indispensable for change.

NOTE

1. David Shearman, "Kyoto: One Tiny Step for Humanity," Online Opinion, March 4, 2005, at <http://www.onlineopinion.com.au/view.asp?article=3085>.

Acknowledgments

The authors thank Praeger and in particular Hilary Claggett for her patience and guidance. It has been a pleasure for us to work with her.

— 1 —
Has Democracy Failed?

A democracy is a government in the hands of men of low birth, no property, and vulgar employments.

—*Aristotle*

DEMOCRACY AND THE MODERN WORLD

Democracy has become the Holy Grail of Western culture. It is preached in almost biblical terms. President George W. Bush hands out his favorite book, *The Case for Democracy* by Natan Sharansky,[1] to fellow statesmen and reminds us that "democracy means freedom and prosperity." It is easy to see why President Bush is inspired by Sharansky, the first political prisoner to be released by Gorbachev in 1986. Sharansky spent nine years in prison for fighting for the rights of Soviet Jews and rightly enough appreciates the value of personal freedom. He was courageous. He has valuable insights into the Middle East peace process, but his celebration of the value of Western democracy is, we believe, naïve, and he has yet to reach the insightful disillusionment of Aleksandr Solzhenitsyn. In his argument for democracy, Sharansky is concerned only with issues of personal liberty. He doesn't move beyond this individualist perspective to consider the major threats to the continuation of human life and civilization. He epitomizes the failure of supporters of liberal democracy to grasp the significance of the global environmental crisis.

We will confront the reader with problems of such magnitude that issues of personal liberty pale into insignificance. We present the case against democracy, showing how freedom and liberalism have the potential to propagate environmental tyranny far greater than any threat posed by the former Soviet Union. The future fruits of liberal democracy may prove to be more bitter than even the gulags of the Soviet system, as horrible as the gulags were.

Let us be clear about one point from the very beginning of this text. The authors are not living fossil Marxists attempting to rehabilitate the Soviet regime. We agree that existing authoritarian societies, largely based upon Marxist doctrines, have had an appalling environmental record. We accept that there is no example of an existing authoritarian government that does not have a record of environmental abuse. We also accept that all existing authoritarian governments have a worse environmental record than all liberal democratic societies. Being "least worst" of a bad bunch is not a logically good argument for the acceptability of the "least worst" option. As a matter of rational argument, defenders of liberal democracy must be forced to do better than merely ignore the long existing problems of democracy, first noted by Plato (427–347 B.C.). We contend that there are other forms of authoritarian government beyond the failed Marxist version. We discuss a Platonic form of authoritarianism based upon the rule of scientific experts, and, as we detail in chapter 8, this hypothetical system is not based upon Marxist principles. We are critics, on ecological grounds, of the capitalist economic system and existing authoritarian systems. We argue that even the allegedly more environmentally preferable liberal democratic societies fail to provide humanity with ecologically sustainable structures. We accept that mention of authoritarian government will horrify the reader with visions of dictators who have strutted during the past century, but we remind that many have been elected under democratic systems.

Attributes of democracy are recognized as questionable even by democracy's defenders. George Monbiot has said, "It is the unhappy lot of humankind that an attempt to develop a least-worst system (of governance) emerges as the highest ideal for which we can strive."[2] There are two positives identified by Monbiot. Democracy is the only system with potential for its own improvement without internal violence, and democracy has the potential to be politically engaging for citizens. Of course, democracy confers the freedom to choose; political systems other than liberalism and authoritarian systems of governance may be selected. We will accept that democratic systems may have advantages, such as self-correctability. However the existence of some merits and some virtues does not show that the system as a whole is satisfactory or sustainable in the long term. There are other independent reasons and arguments for the rejection of democracy, and, as we will argue, liberalism as well.

From altruistic beginnings, however, liberal democracy and its institutions have become a mechanism for powerful nations to control the world by commercial invasion, and sometimes crusading wars are launched to deliver it to the nonbelievers. Like communism, it can be evil. Those living in a liberal democracy are cushioned from these happenings and their consequences. Our freedom of body and mind, and the abundance of materialism, offer a pleasant life that can be rejected only with difficulty, for what is the alternative? Thus democracy has to be defended and those not with us are against us in our quest to maintain our lifestyle. Those attacking or even criticizing democracy will be seen as enemies, for the benefits of materialism are at stake. These material considerations have usurped the theoretical attributes of democracy.

Not only are the democratic citizens cushioned by the comfort of consumerism, but they are manipulated by the psychological apparatus of government, a politics of fear that confers unlimited power to enact legislation and wars never put to democratic test. In his Nobel Prize for Literature acceptance speech, Harold Pinter referred to "a clinical manipulation of power while masquerading as a force for good. It's a brilliant, even witty, act of hypnosis."[3] Pinter was referring to the United States, but his words are applicable to most of the Western democracies. Referring to truth, Pinter warns that the search for truth must never stop and it cannot be adjourned or postponed, and objectivity is essential:

Political language, as used by politicians, does not venture into any of this territory, since the majority of politicians, on the evidence available to us are interested not in truth but in power and in the maintenance of that power. To maintain that power it is essential that people remain in ignorance, that they live in ignorance of the truth, even the truth of their own lives. What surrounds us therefore is a vast tapestry of lies, upon which we feed.[4]

This edifice of deceit is surrounded by a shell of spin and bureaucracies and scientists chosen for their compliance. Thinking citizens accept their impotence to influence events, and the professional image of politicians ranks at the lowest along with used car salesmen.

However putting these subjective assessments aside, our analysis of the performance of democracy is diagnostic, using science and philosophy to define the ills. Society can then move forward to discuss the remedies. We will ask, what is the true record of democracy in addressing and preventing the major issues besetting humanity today, such as war, equity, and especially environmental damage? The most important question of our time is whether the democratic system is able to grasp and remedy the emerging ecological crisis facing the entire human race. What is the precise role of liberal democracy in causing this crisis? What is its performance in remediation during the past two or three decades of increasing scientific evidence of the crisis? To further this task, several critical environmental issues will be analyzed. Many failures are diagnosed and in each instance causation is identified as the modus operandi of liberal democracy. We therefore question whether democracy can be modified or reformed to address these problems before they have become irreversible. And if not, how can humanity be governed? We argue that humanity will have to trade its liberty to live as it wishes in favor of a system where survival is paramount. Perhaps this choice should not be put for democratic approval, or humanity will elect to live as it wishes.

There is also another important point that will recur in our argument, but which requires emphasis now to avoid unnecessary confusion. In a book about democracy it is prima facie reasonable to expect a definition of "democracy": "democracy is X." Defenders of democracy have a problem in saying what "X" actually is. There are a multitude of definitions of democracy and to attempt to taxonomize now would be distracting from this overview. Further, we contend that democracy is conceptually incoherent, in some of its versions at least. Thus one of the problems of democracy is that there is no universally accepted definition that can be worked into an introductory chapter without immediately raising philosophical issues of contention. As we wish to develop an ecological critique of democracy in all its forms and a philosophical rejection of democracy per se, we are not disturbed by not being able to offer the reader an initial, simple definition. There are in our opinion no such satisfactory definitions, for all such definitions (e.g., government of the people, by the people, for the people) are even vaguer and less informative than the concept of democracy, as we show in chapter 5. For the moment we invite the reader to operate with her or his own intuitive understanding of democracy, and in chapter 5 we will criticize the standard accounts. In chapter 7 we will also reject liberalism as a philosophical position.

For the purposes of developing an ecological critique of democracy it is first necessary to understand the basis of the environmental crisis facing humanity. Almost all environmental writers blame the crisis on liberal capitalism. We argue that even if liberal capitalism ceased to exist there would still be the potential for an environmental crisis because of the destructive tendencies within the heart of democracy itself.

CRISIS? WHAT CRISIS?

This impending crisis is caused by the accelerating damage to the natural environment on which humans depend for their survival. This is not to deny that there are other means that may bring catastrophe upon the earth. John Gray for example[5] argues that destructive war is inevitable as nations become locked into the struggle for diminishing resources. Indeed, Gray believes that war is caused by the same instinctual behavior that we discuss in relation to environmental destruction. Gray regards population increases, environmental degradation, and misuse of technology as part of the inevitability of war. War may be inevitable but it is unpredictable in time and place, whereas environmental degradation is relentless and has progressively received increasing scientific evidence. Humanity has a record of doomsayers, most invariably wrong, which has brought a justifiable immunity to their utterances. Warnings were present in *The Tales of Ovid* and in the Old and New Testaments of the Bible, and in more recent times some of the predictions from Thomas Malthus and from the Club of Rome in 1972, together with the "population bomb" of Paul Ehrlich, have not eventuated. The frequent apocalyptic predictions from the environmental movement are unpopular and have been vigorously attacked.

So it must be asked, what is different about the present warnings? As one example, when Sir David King, chief scientist of the UK government, states that "in my view, climate change is the most severe problem that we are facing today, more serious than the threat of terrorism,"[6] how is this and other recent statements different from previous discredited prognostications? Firstly, they are based on the most detailed and compelling science produced with the same scientific rigor that has seen humans travel to the moon and create worldwide communication systems. Secondly, this science embraces a range of disciplines of ecology, epidemiology, climatology, marine and fresh water science, agricultural science, and many more, all of which agree on the nature and severity of the problems. Thirdly, there is virtual unanimity of thousands of scientists on the grave nature of these problems. Only a handful of skeptics remain.

During the past decade many distinguished scientists, including numerous Nobel Laureates, have warned that humanity has perhaps one or two generations to act to avoid global ecological catastrophe. As but one example of this multidimensional problem, the Intergovernmental Panel on Climate Change (IPCC) has warned that global warming caused by fossil fuel consumption may be accelerating.[7] Yet climate change is but one of a host of interrelated environmental problems that threaten humanity. The authors have seen the veils fall from the eyes of many scientists when they examine all the scientific literature. They become advocates for a fundamental change in society. The frequent proud statements on economic growth by treasurers and chancellors of the exchequer instill in many scientists an immediate sense of danger, for humanity has moved one step closer to doom.

Science underpins the success of our technological and comfortable society. Who are the thousands of scientists who issue the warnings we choose to ignore? In 1992 the Royal Society of London and the U.S. National Academy of Sciences issued a joint statement, *Population Growth, Resource Consumption and a Sustainable World*,[8] pointing out that the environmental changes affecting the planet may irreversibly damage the earth's capacity to maintain life and that humanity's own efforts to achieve satisfactory living conditions were threatened by environmental deterioration. Since 1992 many more statements by world scientific organizations have been issued.[9] These substantiated that most environmental systems are suffering from critical stress and that the developed countries are the main culprits. It was necessary to make a transition to economies that provide increased human welfare and less consumption of energy and materials. It seems inconceivable that the consensus view of all these scientists could be wrong. There have been numerous international conferences of governments, industry groups, and environmental groups to discuss the problems and develop strategy, yet widespread deterioration of the environment accelerates. What is the evidence?

The Guide to World Resources, 2000–2001: People and Ecosystems, The Fraying Web of Life[10] was a joint report of the United Nations Development Program, the United Nations Environment Program, the World Bank, and the World Resources Institute. The state of the world's agricultural, coastal forest, freshwater, and grassland ecosystems were analyzed using 23 criteria such as food production, water quantity, and biodiversity. Eighteen of the criteria were decreasing, and one had increased (fiber production, because of the destruction of forests). The report card on the remaining four criteria was mixed or there was insufficient data to make a judgment.

In 2005, *The Millennium Ecosystem Assessment Synthesis Report* by 1,360 scientific experts from 95 countries was released.[11] It stated that approximately 60 percent of the ecosystem services that support life on earth—such as fresh water, fisheries, and the regulation of air, water, and climate—are being degraded or used unsustainably. As a result the Millennium Goals agreed to by the UN in 2000 for addressing poverty and hunger will not be met and human well-being will be seriously affected.

THE ENVIRONMENT IN INTENSIVE CARE

The responsibilities and performance of the liberal democracies in these scenarios will be analyzed using our training in medicine, science, the law, philosophy, and social science. There are so many variables perpetrating today's problems that reductionism cannot hope to offer analysis and solutions. Knowledge and understanding has to be global and multidisciplinary. Each human life depends upon the integrated function of heart, lungs, brain, liver, kidneys, nerves, and muscles to constitute as an integrated ecological system that forms one human body. It is useful to regard the living earth in this way, as a complex integration of interdependent systems to form one planet.[12] Just as we document and recognize environmental damage, we can assess the living earth like a patient as healthy or ill. The documentation of environmental damage indicates that the earth is ill. But worse, there is evidence that this patient is already in the intensive care unit, for several of its organs are failing. "Multiorgan failure" is written in the patient's records. In such situations the outcome cannot be predicted. Unfortunately ecological and medical science cannot tell us whether the human body or the ecological system has reached the point of irretrievable collapse. In a paper published in the Journal *Nature*,[13] studies on deforestation, endangered species, and eutrophication (when water is choked by the presence of too many nutrients) of lakes all showed resistance of these systems to gradual environmental damage, and then sudden collapse took place without warning. Collapse means demise of an ecological system that is of service to humanity.

Can we draw lessons from the human patient in the intensive care unit? The patient's resuscitation is in the hands of a leader, the expert doctor in intensive care, and a team of nurses and scientists, which combines leadership with expert knowledge, decision making, speed, dedication, and compassion. The leader does not explore the public opinion polls to see what can be tolerated or is popular. He or she does not act to preserve their position at the next election and is not influenced by corporatism or the perceived state of the economy. There is one collective, unsullied goal, to recognize the emergency, to make a skilled diagnosis based upon scientific assessment and to restore health before the situation becomes irreversible. This physician uses the precautionary principle by taking action to support each organ to the full in case collapse is impending. Experience suggests that a human health crisis is best dealt with in this way. When the patient is the living earth, we will ask whether the institutions of liberal democracy and liberal capitalism measure up to the task. Viewed in the light of intensive care medical metaphor, we can also ask whether decision-making structures per se are the appropriate mechanisms when it is the biosphere itself that is in intensive care.

Is there a crisis? To answer this question we have analyzed several key indicators necessary for the survival of human civilization. In assessing the adequacy and sustainability of these indicators we have been mindful of the expected increase in the world's population of 6 billion to at least 9 billion by 2050, a figure that has wide acceptance as a likely outcome. We have chosen to study supplies of fresh water for there is a finite volume of water that falls on the earth. Fresh water supplies are already inadequate for the basic needs of many sections of the world's population. We have examined the sustainability of fish stocks as a measure of food supply, though we could have chosen cereals or other foods. The harvesting of fish is probably at its peak and many species of fish will not recover from overfishing. Recognizing that civilization cannot exist in its present form without ecological services, we have studied biodiversity. Large extinctions of species are already occurring, and this trend will accelerate with global warming. We have analyzed the data on climate change. Here there is evidence from many scientific sources that warming is occurring, and, unless greenhouse emissions are controlled, the future of living things that support our own survival is dim.

We have also examined the consumption of fossil fuels because their reckless use is the inherent cause of climate change and also because it is predicted that the depletion of oil sometime in the next few decades will severely reduce the number of people who can be fed. This is because oil has been the fundamental resource for fertilizers, mechanized farming, and transport that has supported the world's burgeoning population. There are carefully researched predictions that a world without oil can support only 2 billion people.[14]

It is fair to say that whichever environmental parameter is being assessed, there is a remorseless deterioration. Degradation is the express train, remediation is the slow train, stopping and starting and never catching up. For each of these examples of environmental deterioration, the role of liberal democracies in causation will be analyzed. Everyday decisions are made to delay the slow train even further and, although the thinking behind these decisions often has a simple explanation, in reality it is a complex decision based on values and cultural, political, and corporate influence. In December 2004, the European Community made a decision on fishing quotas for member states. Scientific data indicates that the depletion of cod stocks in northern waters was at the point of collapse, and there is serious doubt that they will recover. There was strong scientific advice that exclusion zones must be established in the hope of recovery. The political representatives of these wealthy, well-fed, liberal democracies exercised what they saw as their democratic mandate and severely curtailed this proposal in the interest of "jobs now."[15] Every day in liberal democracies, countless decisions like this one remorselessly eat away at the environment.

In the same month, the UK assessed its performance in reducing greenhouse emissions. Of all the leaders in the world, Tony Blair has grasped the implications of global warming caused by greenhouse emissions.

He recognized that the Kyoto proposal for the industrialized world to reduce emissions to 5 percent below 1990 levels by 2012 was tokenism compared to the 60 to 80 percent advised by the scientific community. The UK set a target of reducing emissions to 20 percent below 1990 levels by 2010. This target will be missed, though the Kyoto target will be achieved mainly because of the UK's reduction in the use of coal, due to the previous closure of mines. If the world's leading advocate for reducing emissions, one who has the maximum parliamentary power that democracy can bestow, cannot deliver environmental outcomes, what hope is there for other countries to reduce emissions? Mr. Blair has not been able to speed up the slow train but more importantly his fast train is increasing its speed. The nameplate proudly displayed on the engine is "The Growth Economy." Admittedly the reasons for failure are more complex than this and include human factors of psychology and denial and the self-interest of the major influences on liberal democracy. These are discussed in chapter 6.

"THE COMMONS": A STATE OF DENIAL

Our innate responses must be understood, for they have a profound influence on our ability to respond to the ecological crisis. Humans are born with psychological mechanisms that significantly influence behavior. Richard Dawkins points out that "if you wish to build a society in which individuals co-operate generously and unselfishly towards a common good you can expect little help from biological nature."[16] Thus, self-preservation and the need to procreate determine our quest for goods, status, and power. Humanity's inability to think long term is related to the brain having hard wiring from our "paleolithic heritage."[17] Over hundreds of millennia we had to adapt to the conditions of a local environment. We had to think short term with an emotional commitment to the limited space around us and to a limited band of kinsmen. This is the Darwinian priority of short-term gain that bestowed longevity and more offspring upon a cooperative group of relatives and friends. As a result, we ignore any distant possibility not yet requiring examination. Global warming and loss of ecological services are seen only as distant possibilities. Families cannot comprehend responsibility beyond their grandchildren, and in Western societies the increasing number of couples without children tends to limit responsibility to their own lifetime. Indeed, Western society has moved increasingly to the delivery of short-term needs, solutions, and profit and a disregard of anything that is not self-centered. One of the authors, on asking medical colleagues how they feel about the effect of climate change on the future lives of their children, has found that a common response is "that's their problem."

The state of denial is relevant to the present discussion. When we are faced with a problem that extends beyond our local environment, or when it involves distant individuals and races, then it is not of relevance to our needs. The defensive mechanism of denial is activated. The psychology of denial has been studied in relation to human rights and to poverty and famine.[18] Denial often relates to the enormity of the problem because one individual can do little about it. An individual can accept the scientific evidence of, say, climate change, but deny responsibility or blame others for creating the problem. The provision of more information may increase denial and lead to antagonism to the cause. Images of starving children are suppressed and requests for donations ignored. Denial is the basis of the language used to describe the unpalatable problems. In war, mass murder becomes "cleansing," and with global warming the expected inundation and drowning of Pacific islanders by tidal surges are described by politicians and governments as a "human impact."[19]

There is a further human factor that requires discussion: religion. Whilst it is possible that the environmental nihilism of President Bush may be due to denial, there is a much more worrying possibility, that his religious beliefs may be responsible. There are more than 200 Republican legislators in the U.S. government who are Christian fundamentalists, many of whom belong to sects that believe the future of the planet is irrelevant because it has no future.[20] They are living in the "end times," after which the son of God will return. Environmental destruction is to be welcomed, even hastened for it is a sign of the coming Apocalypse when they will enter heaven and the sinners will suffer eternal hellfire. One of these fundamentalists, Senator James Inhofe, chairs the powerful Senate Environment and Public Works Committee, and as a member of the Bush regime he has helped curtail many important environmental controls such as laws on clean air, clean water, endangered species, pollution limits for ozone, car emissions, coal-fired power stations, mercury, and many more. An analysis of these actions, in conjunction with the President's unguarded statements such as his use of the word "crusade" for his actions in Muslim countries, suggests that decisions are not made on the basis of rational thought or science, but on the tenets of religious fundamentalism. Our discussion of the problems of democracy in a warming world will therefore include the positive and negative role of religions.

In an important scientific paper published in 1968 entitled "The Tragedy of the Commons,"[21] Garrett Hardin discussed a number of problems for which he believed there was no technical solution. The problems in question required a change of values for their solution. He hypothesized that the population problem was such a "no-technical solution" problem. In the course of his argument he introduced the idea of the tragedy of the commons. The individual pursuit of self-interest will lead rational neoclassical economic agents to exploit a resource to extinction: as all such agents act in this way, a commons such as the ocean or the atmosphere will become degraded. Individual self-interest can lead to collective environmental disaster. We argue that liberal democracy is ecologically flawed as a social system because it leads to the tragedy of the commons. Fifty one percent of the people can vote to destroy a resource (or simply act to maintain unsustainable lifestyles), which 49 percent of the people wish to preserve. There is thus the potential for ecological destruction existing in the heart of democratic institutions. We will return to this point with specific examples. The fundamentals of our critique of liberal democracy and democracy itself are now discussed.[22]

LIBERAL DEMOCRACY

Liberal democracy is clothed in a long overcoat, but what is underneath is revealing, as we will see. "Liberal democracy" has become a sweet sounding word of the advertiser; it caresses the mind like the name of a perfume or a succulent chocolate. It is perhaps surprising that (to use George Orwell's terminology) this Old-speak name,[23] liberal democracy, has not been replaced by a brand name, perhaps "imagine," for it conjures freedom, success, and prosperity, or at least the prospect of these goals for the poor and oppressed. Once "imagine" is the "Newspeak," then the populace can be united even more fervently for more. The slick advertisers and spinners are the tools of powerful governments of the West who implement the creed of liberal democracy, usually with missionary zeal and financial power, but occasionally with covert or overt force. A messianic United States is the leader of this cultural movement that prescribes the primacy of the market, human rights, and personal ownership as principles that must not be transgressed. How did this eventuate?

Modern democracy was born over two centuries ago, a child of the rapidly developing industrialization, commerce, and trade of North America and Europe. Indeed it grew up in symbiosis with capitalism and is now inseparable. Democracy provides the freedom of action, the liberalism, for each individual not only to fulfill all material needs but to accumulate unlimited wealth and the commercial power it confers—at least in principle. This power and influence stems from the dependence of all democracies on the mantra of economic growth so necessary to provide the employment and consumer goods to satisfy and placate the peoples.

Liberalism is the belief system that holds that the freedom of the individual, especially in economic matters such as trade and labor relationships, is of principal importance. The liberal sees society made up of individuals like a house is made up of bricks to create an inanimate structure. Margaret Thatcher said it all in her statement that "there is no such thing as society," that is, only individuals exist.[24] Thus society is not a complex system, or whole, like the human body. Christianity had, in some respects, anticipated this idea that all humans were raceless, placeless (but not sexless) individuals who had souls and stood equally before a judging God. Christianity carried this seed in its womb to give birth to Protestant American democracy. Liberalism had borrowed this individualism and replaced it with a secular version that gave the individual laborer and consumer a new place before the new God—the market.

It was necessary for the rising class of capitalists and merchants who slowly began to challenge the feudal order and power of the Roman Catholic Church to have a set of beliefs that legitimized their new world order. The traditional feudal Church saw people as having a God-given fixed station in life. The Church also opposed usury—the lending of money at a rate of interest. It was a sin to earn money whilst you were asleep. These were beliefs that placed limits upon trade and commerce—that is, money making. For the emerging capitalist class, these restraining beliefs had to be undermined. And indeed, in time, these beliefs were challenged and replaced by a new philosophy, liberalism. Individuals were now said to be *free* in the sense of no longer being bound to a feudal master. Instead, they were free to sell their labor on the market—or free to starve. In this book we expose the mythical legitimizing role that liberal democracy gives to the capitalist social order. We reject the myth that feudalism was a dark oppressive system, whereas liberal democracy was and is a force of light, salvation, and emancipation. On the contrary, there is much evidence to suggest that liberal democracy—the meshing of liberalism and democracy—is the core ideology responsible for the environmental crisis. Liberal democracy, it should be noted, although in principle as a matter of logic is conceptually distinguishable from capitalism, has become a matter of real politics intrinsically enmeshed with capitalism, and it is virtually impossible to separate the effects of each.

Modern democracy, the idea that government should be by the "will of the people" (whatever that means) is conceptually linked to the notion of liberalism. It is inconceivable that there could arise, historically, a system that gave primacy to an *individual's* vote that did not justify or base this belief system on a philosophy that saw the individual, rather than society, as an "organic whole" as being of primary value.[25] Liberalism thus supplied the philosophical justification for democracy (and capitalism), just as Christianity had earlier supplied the religious justification for the rule of kings and queens through the doctrine of divine right. The secular version of this divine right has become market forces that in many ways now rule our lives more oppressively and totally than any king or dictator in the premodern world.

An evolutionary and therefore genetic mechanism relevant to our analysis is the need and acceptance of authoritarian social structures conferred upon us by our primate ancestors. These forces can even be seen to operate within a liberal democracy in which leaders and democratic institutions themselves gradually evolve to become more authoritarian. Freedom and individuality expressed through the market economy result in elites widening the gap between rich and poor and enriching themselves by acquisitions in developing countries under the guise of freedom and democracy. Maladaptations of society as defined by Stephen Boyden[26] become more common, for example the economic view that retail spending is good for society or the accumulation of vast assets by the rich that they cannot possibly use or spend in their lifetimes. The number of billionaires in the world is increasing rapidly and the majority are in the liberal democracies. As we will see in the discussion to follow, many liberal democracies are moving visibly toward authoritarianism. Governments see this as an option to protect their power, and many of their rich supporters favor it to protect their assets.

It will be argued in chapter 6 that liberal democracies are inherently unstable and move slowly but surely to authoritarianism. Theorists who have seen liberal democracy as representing humanity's final political system have adopted a too narrow historical perspective, which can be corrected by adopting a biohistorical or sociobiological view of the human species. We should not be blind to the possibility that an authoritarian meritocracy might have advantages in world crisis management compared to the present democratic mediocracy. Our patient in the intensive care unit could not be managed successfully under liberal democracy. Recognizing that totalitarian states have caused as much, if not more, environmental damage as the liberal democracies, we will nevertheless argue in chapter 4 that some historical totalitarian regimes have averted some catastrophic environmental damage by dictate.

We will document the personal and democratic failures that render the environmental crisis difficult to address. An altruistic, able, authoritarian leader, versed in science and personal skills, might be able to overcome them. But liberal democracy predisposes the election of the slick wielders of the political knife and then encumbers them with the burdens of economic chains and powerful self-interested corporates who cannot be denied. They fuel the growth economy that preserves their power and that of government. It is instructive to ask our democratically elected leaders: What do you see as the endpoint of this liberalized growth economy? Surely to maintain this growth to infinity is unsustainable? Yet this growth is necessary for the present economic system to survive and satisfy the perceived material needs of humanity. Our leaders cannot provide an answer to this question. To some it falls beyond their elected period, and they do not have to address it. To others there is the hope that science and technology will capture the carbon dioxide of climate change, create hydrogen fuel from water, and feed the millions with genetically modified foods. But in general it is not an issue that democratic societies are addressing in a way that will encourage solutions.

DEMOCRACY AND THE POWER BROKERS

It is possible to see the control of society firmly grasped by a brotherhood that resembles a biological ecological system. Like the soil, the forest, or the coral reef, its strength lies in mutual support and interdependence of all organisms and components. The web of power and profit embraces the market, the banks and financial institutions, regulators (national and international), the liberal democracies, the press, the media and advertising industries, and the military industrial complex. The governments espousing liberal democracy are but the compliant arms and hands of the system. They provide the human fodder from their universities. They retain power by servitude. As we will show in chapters 6 and 10, those at the top of the food chain are the corporations. They operate for profit alone, protected by law that absolves them from other responsibility. Their leaders, who live a double life of family care and principle at home, but plunder the world for gain, are the conquistadors of today. Like the Spanish noblemen, the Chief Executive Officers have become the pillars of society. The spoil is no longer gold, but black gold (oil), plantations, and water industries. They would not recognize themselves as the ecology of evil, but for the future of the world's environment that's what history may judge them as. For some, such as Clive Hamilton in *The Disappointment of Liberalism and the Quest for Inner Freedom,*[27] the source of our difficulties lies not in democracy itself but in its undermining by lobbyists who act for corporatism and the market. Liberal capitalism, not liberal democracy, is the real culprit. These thoughts are echoed by George Monbiot:

Meaningful action on climate change has been prohibited by totalitarian capitalism. When I use this term I don't mean that the people who challenge it are rounded up and sent to break rocks in Siberia. I mean that it intrudes into every corner of our lives, governs every social relation, becomes the lens through which every issue must be seen. It is the total system which leaves no molecule of earth or air uncosted and unsold.[28]

Surely Hamilton and Monbiot fail to understand the strength and complexity of this ecological system of evil into which democracy has descended. Democracy is but a cog in this juggernaut causing environmental degradation. Liberal capitalism and democracy have fused together. Liberal capitalism, the retrovirus, has become part of the genetic material of democracy and is directing the enterprise. It is not just an imperfection that can be corrected without dismantling this relationship. As we will demonstrate, these colossal environmental problems, both existing and impending, have been accelerated by the freedoms and corruption of democracy and are unlikely to be solved by this system of governance. Thus we agree with the well-known critique from left-environmental writers that the primary cause of the environmental crisis is the existence of an ecologically unsustainable economic system, capitalism. However we go further than these critics in implicating liberal democracy and democracy in general in causing this environmental crisis and specifically preventing its solution. For a variety of reasons, detailed by us, democratic institutions are not suited to deal with crisis care situations. If you needed to have major heart surgery you would not wish your operation coordinated by a democratically elected team of surgeons. With respect to liberal capitalism, in chapter 10 we come to the same conclusions as John Perkins in *Confessions of an Economic Hit Man*.[29] Perkins worked for the covert U.S. National Security Agency. He has said, "We build a global empire. We are an elite group of men and women who utilize international financial organizations to foment conditions that make other nations subservient to the 'corporatocracy' running our biggest corporations, our government and our banks. The subservience is financial and the government is that of the USA."[30] Liberal capitalism, we will argue, is a force acting to produce an authoritarian rule by corporate elites. Although enmeshed with liberal democracy its ultimate goals are antagonistic to it, and in the long term act to undermine it.

We predict that democracy, like communism, will be but a moment in human history. Its transformation into authoritarian rule is likely to be catalyzed by its failure to deliver solutions to the environmental crisis. We can speculate on the preferred form of authoritarianism and in chapter 9, "Plato's Revenge," we define the essential ingredients. We can wish for the intensive care model, but we are unlikely to be so fortunate. However, a consideration of the form of social cohesion necessary to maintain civilization in a no-growth economy is vital, for this is where we must go for survival. A new religion or perhaps spirituality to replace the market and consumerism will necessarily embrace the earth and all its sacred life.

To ask where liberal democracy is leading us is not a welcome question, as the liberalism conferred by democracy is the linchpin of our culture.

Therefore simply asking the question leads to a response: What other system of governance and economics can we turn to? Are we to return to living in caves? Are we to return to the inefficiencies of socialism, the iniquities of communism, or the cruelties of fascism? But the question must be asked because our present culture is instrumental in directing us to environmental change that is likely to devastate our civilization during this century. There are those such as Jonathon Porritt who contend that capitalism, which has brought humanity to this parlous condition, can nevertheless deliver the solutions.[31] We disagree and will explain our reasons.

When we review the alarming data about accelerating climate change, it is our duty to place in the mind of the reader the magnitude of the response required from humanity. To ancient Egypt, it might be the equivalent of the building of the pyramids, a task that seems superhuman even today. To civilization today we need to think of resolve for a technological revolution as vast as the Manhattan project and NASA's space endeavors, proceeding not just in the United States but in all developed countries and delivered with the vision and acceptance of a Marshall Plan. But even more vital, a revolution in lifestyle, a new paradigm delivered to the populace with the flair of Marshall and with the authoritarian brilliance of Napoleon, who revised the chaotic French legal system overnight and imposed it in the morning. Today, it is debatable whether we can wait for democratic reform, bit by bit, election by election, and decade by decade. It is our task in this book to provide the evidence.

NOTES

1. Natan Sharansky, *The Power of Freedom to Overcome Tyranny and Terror* (Public Affairs, New York, 2004).

2. George Monbiot, *The Age of Consent: A Manifesto for a New World Order* (Flamingo, London, 2003), p. 41.

3. Harold Pinter Nobel Lecture, "Art Truth and Politics," 2005, at <http://nobelprize.org/literature/laureates/2005/pinter-lecture-e.html>.

4. Ibid.

5. John Gray, *Al Qaeda and What It Means To Be Modern* (Faber and Faber, London, 2003).

6. David A. King, "Climate Change Science: Adapt, Migrate or Ignore?" *Science,* vol. 303, no. 5655, January 9, 2004, pp. 176–177.

7. Intergovernmental Panel on Climate Change, *Climate Change 2007: Fourth Assessment Report,* at <http://www.ipcc.ch/>.

8. Joint Statement of the Royal Society of London and U.S. National Academy of Sciences, *Population Growth, Resource Management, and a Sustainable World,* 1992, at <http://www.dieoff.com/page7.htm>. According to this statement, "the future of our planet is in the balance. Sustainable development can be achieved but only if irreversible degradation of the environment can be halted in time."

9. Ibid.; U.S. National Academy of Sciences, *Joint Statement by 58 of the World's Scientific Academies,* at <http://dieoff.org/page75.htm>; Joint Resolution of the U.S. National Academy

of Sciences and Royal Society of London, *Towards Sustainable Consumption,* 1997, at <http://www.royalsoc.ac.uk/document.asp?tip = O&ID = 1907>; World Scientists' Call for Action, Union of Concerned Scientists, December 1997, at <http://go.ucsusa.org/ucs/about/page.cfm?pageID = 1007>; Inter-Academy Panel, May 2000, *Transition to Sustainability in the 21st Century: The Contribution of Science and Technology,* at <http://www.interacademies.net/cms/about/3143/3552.aspx>; World Resources Institute, *Guide to World Resources, 2000–2001: People and Ecosystems; The Fraying Web of Life,* April 2000, at <http://pubs.wri.org/pubs_description.cfm?PubID = 3027>.

10. World Resources Institute, *Guide to World Resources, 2000–2001,* from note 9.

11. Millennium Ecosystem Assessment, *Millennium Ecosystem Assessment Synthesis Report,* March 2005, at <http://www.millenniumassessment.org/en/Synthesis.aspx>.

12. James Lovelock, *Gaia: A New Look at Life on Earth* (Oxford University Press, Oxford, 1979).

13. Marten Schaffer, et al., "Catastrophic Shifts in Ecosystems," *Nature,* vol. 413, October 11, 2001, pp. 591–596.

14. Russell Hopfenberg and David Pimentel, "Human Population Numbers as a Function of Food Supply," *Environment, Development and Sustainability,* vol. 3, no. 1, 2001, pp. 1–15.

15. Andrew Osborn, "Fishing Grounds Escape Closure Threat," *The Guardian,* December 22, 2004, at <http://www.guardian.co.uk/print/0,,5091026–106710,00.html>.

16. Richard Dawkins, *The Selfish Gene* (Oxford University Press, Oxford and New York, 1989).

17. E.O. Wilson, *The Future of Life* (Little, Brown, London, 2002).

18. S. Cohen, *States of Denial: Knowledge About Atrocities and Suffering* (Polity Press, Cambridge, 2001).

19. David Shearman, "Time and Tide Wait for No Man," *British Medical Journal,* vol. 325, 2002, pp. 1466–1468.

20. Glenn Scherer, "The Godly Must be Crazy," *The Grist Magazine,* October 2, 2004.

21. Garrett Hardin, "The Tragedy of the Commons," *Science,* vol. 162, 1968, pp. 1243–1248.

22. Democracy is a term derived from the Greek, *demos,* the "people." In 593 B.C. a Council of Four Hundred was elected in Athens from the parishes *(demes)* of each tribe. The aristocracy, based on the ownership of land, had changed to embrace the commercial interests of trade and commerce and the influx of artisans and workers attracted by these activities. Tribal structures were weakened and civilization's first rudimentary democracy evolved. Greek democracy was but a moment in Greek history; it passed as we believe modern democracy will also pass. Democracy is not a stable long-term state for the human species, or so we will argue in this book.

23. George Orwell, *Nineteen Eighty-Four* (Penguin Books, Middlesex, England, 1954).

24. Interview for *Women's Own,* Thatcher Archive: COI Transcript, October 31 1987, at <http://www.margaretthatcher.org/speeches/displaydocument.asp?docid=106689>.

25. Today of course there are illiberal democracies such as Singapore, which do not have a free society based upon individualism, but that still have democratic elections. Such societies have had a democratic electoral system artificially grafted onto an authoritarian society through British colonialism. Illiberal democracies are only possible because of the prior historical emergence of liberal democracies.

26. Stephen Boyden, *The Biology of Civilisation: Understanding Human Culture as a Force in Nature* (University of New South Wales Press Ltd, Sydney, 2004).

27. Clive Hamilton, *The Disappointment of Liberalism and the Quest for Inner Freedom* (Australia Institute Discussion Paper Number 70, Canberra, 2004).

28. George Monbiot, "Awareness is Not Enough," *Guardian Weekly,* July 22–28, 2005.

29. John Perkins, *Confessions of an Economic Hit Man* (Berrett-Koehler Publishers Ltd, San Francisco, 2004).

30. Ibid., p. xvii.

31. Jonathon Porritt, *Capitalism As If the World Matters* (Earthscan, London, 2006).

— 2 —

Time and Tide Wait for No Man

> The truth about the climate crisis is an inconvenient one that means
> we are going to have to change the way we live our lives.
>
> —*Al Gore*

A WARMING PLANET

Climate change is bringing slow, sometimes hidden, damage to the earth's living systems. There are some marches in the streets demanding action to reduce greenhouse emissions, but there will no doubt be marches protesting the consequences of drastic measures to control emissions—for example, increases in fuel taxes—if they are ever implemented. Humanity's horizons are limited. There is no grieving at the extinction of the dodo, for it was not present in our lifetime. There is little concern for the predicted demise of one quarter of the world's species during this century. Most will remain during our lifetime. Ambivalence abounds, for the spring of northern climes bursts forth early and life is more equitable for some. Those with conviction are the Inuit whose life is melting away beneath them and the South Sea islanders whose homelands are being submerged. They do not have the power or means to march in London or New York to disturb the equanimity and joy of an early spring. Hence it will be a supreme test for democratically elected governments to show the leadership and resolve to control the emerging crisis. To date they are failing.

The perils that await humanity if greenhouse emissions are not controlled quickly will now be described, and the necessity for humanity to move quickly from fossil fuels to alternative energies will be discussed in the context of the inertia of the polluting democracies. For 15 years now there has been an accumulation of scientific data indicating that global warming is having effects observable not only by the scientists but by the general population. To stabilize the climate this century, a reduction of greenhouse emissions of 60 to 80 percent is needed. Signatories of the Kyoto Protocol committed themselves to reduce emissions by 5 percent of 1990 levels by 2012, but this meager reduction is not being achieved by most countries. In a few cases it is being achieved by the massaging of figures. Based on the performance of the past 15 years therefore there can be little hope of preventing global warming before changes to the earth bring chaos to civilization. It is a situation that might be retrieved with human resolve, leadership, and radical changes in values, economies, and lifestyles, but humanity is beset by the moral turpitude of the leaders of the most powerful and innovative nation in the world. The United States, with 5 percent of the world's population, is responsible for a quarter of the world's greenhouse emissions. The misfortune for the environment and the world's peoples is the presidency of George W. Bush, which has sabotaged concerted international action to address the issue.

The average temperature of the earth's surface is maintained at 14°C (57.2°F) by the natural greenhouse effect of the main greenhouse gases in the atmosphere: carbon dioxide, methane, and water vapor. When sunlight reaches the surface of the earth, it is converted into heat, which is then reradiated back into space as infrared radiation. Some of this infrared radiation is absorbed by the greenhouse gases to create a blanket of warm air around the earth. Without these gases, the temperature at the surface of the earth would be −19°C (−2.2°F).[1] Thus the greenhouse effect is a natural phenomenon necessary for the maintenance of life on earth.

There is consensus that the enhanced greenhouse effect is due to the human production of carbon dioxide, methane, nitrous oxide, and halogenated compounds. The greatest contribution to the enhanced effect is from carbon dioxide, which arises from the burning of coal, oil, and gas and from deforestation. The concentration of carbon dioxide in the atmosphere has risen by about 30 percent since preindustrial times and is increasing at 0.4 percent per year.[2] As a result, the average temperature of the earth has risen 0.7°C (1.26°F), which does not seem like much, but it represents a 5 percent rise on the *average* temperature of 14°C (57.2°F).

The atmospheric concentration of carbon dioxide during the past 40,000 years has been determined from glacial ice cores. Its concentration during the Holocene (the current interglacial period) until the start of the Industrial Revolution was stable at 280 parts per million (ppm), but since then has steadily increased and is now over 370 ppm. (It may have risen briefly near the start of the Holocene, but to no more than 330 ppm.) The ice core data have been supplemented in recent decades by direct atmospheric sampling. Modeling, based on different socioeconomic, technological, and climatological scenarios, predicts the concentration of CO_2 by the year 2100 to exceed 490 ppm and to be even as high as 1,260 ppm. Stabilizing the CO_2 concentration even at 450 ppm will require global human-induced carbon dioxide emissions to drop below 1990 levels within a few decades and continue to decrease steadily thereafter. Models show that if it takes a century for humankind to reduce CO_2 emissions to 1990 levels, then a concentration of 650 ppm could result.[3] Overall, the global temperature is expected to increase by 1.4 to 5.8°C (2.52 to 10.44°F) by 2100, with the Fourth Intergovernmental Panel of Climate Change Report believing the rise to be of 4°C (7.2°F), unless society is able to dramatically reorganize itself in order to significantly curtail its greenhouse gas emissions. This rate of temperature increase would be much greater than any during the past 40,000 years.

The scientific data indicating anthropocentric global warming has been accumulated by probably the largest and most rigorous scientific collaboration in history. The Intergovernmental Panel on Climate Change (IPCC) reports have been produced by hundreds of scientists from over a hundred countries. This work is supplemented by studies from many research centers throughout the world. We will summarize the important data to provide a basis for our argument, but for further information the reader is referred to the IPCC report, which has a summary for general readers.[4]

The increase in the global average temperature of 0.7°C (1.26°F) during the twentieth century has already affected many of the earth's biological (see chapter 4) and physical systems. For example, observations have shown a reduction of about two weeks in the annual duration of lake and river ice cover in the midlatitudes and high latitudes of the northern hemisphere during the twentieth century. The extent of northern hemisphere spring and summer sea ice measured by satellites has decreased by 20 percent below average, and the decrease is accelerating.[5] Satellite data have also shown a reduction in 10 percent in snow and ice cover since the late 1960s, and the widespread retreat of mountain glaciers has been well documented globally. Since 1950, observations show a reduction in the frequency of very low temperatures, with a smaller increase in the frequency of extremely high temperatures.[6]

During the twentieth century global sea levels rose between 0.1 and 0.2 meters, and it is very likely that rainfall increased by 0.5 to 1 percent per decade over most midlatitudes and high latitudes of the northern hemisphere.

The frequency and intensity of droughts have been observed to increase in recent decades. The El Niño-Southern Oscillation (ENSO) phenomenon, which affects regional variations of rainfall and temperature over much of the tropics, subtropics, and some midlatitude areas, has shown an increased frequency of warm (El Niño) episodes since the mid-1970s compared with the previous 100 years. Computer models suggest this may be a consequence of human actions.

It is important to recognize that the scientific basis for an understanding of global warming is open to misinterpretation. A small minority of climate change scientists interpret the scientific data differently than the majority. The views of these skeptics have been answered elsewhere.[7] In some cases however misinterpretation is deliberate and is used to delay action by calling for more scientific research. Traditional scientific methods attempt to explore hypotheses or questions using experiments that can give reproducible results. By contrast, the science of global warming is largely based on computer projections, using models validated and improved as theory and current and historical data expand.[8] Forecasts and conclusions must therefore be presented as statistical probabilities. The IPCC uses the words "virtually certain" to mean a greater than 99 percent chance of truth and "very likely" when there is a 90–99 percent chance. Governments, though routinely dealing with considerable uncertainty when it comes to economic forecasts and policy decisions, have yet to fully grasp that uncertainty is also inherent in many fields of physical science. Climate change scenarios, especially if expanded to incorporate the various complex interactions and feedbacks of human society, exemplify this. Governments should adopt the precautionary principle as a basis for policy making, but instead they use uncertainty as an excuse for prevarication or inaction. As a result, humanity is continuing to traverse an increasingly uncertain and risky trajectory that at worst may compromise the sustainability of civilization.

The physical changes described above are predicted to intensify in this century. Warmer temperatures are predicted to increase sea levels by around half a meter, with estimates ranging up to 0.9 meters, due to thermal expansion of seawater and the melting of snow, ice caps, and glaciers. More intense rain is predicted in various parts of the world, and other extreme climatic events, such as storms and droughts are also expected to increase. Some climate models predict monsoonal changes, including increased aridity over large parts of India and Southeast Asia. This could adversely affect food security, lead to increased tension in the nuclear-armed subcontinent, and result in many more ecological refugees.

As pointed out by Ross Gelbspan in *Boiling Point*,[9] progress in acknowledging the issue in the United States has been inhibited by its depiction as solely an environmental issue. In fact it is an issue affecting the entire fabric of society, including food availability from land and sea, fresh water, employment, security, human rights and justice, public health, and the future well-being of humanity. The insurance industry recognizes that the annual increase in financial losses is due to the predicted increases in typhoons and hurricanes. The small island states recognize rising sea levels with inundation of their land and seek legal retribution for the impending economic loss. The poor recognize that they will suffer most from climate change, especially those in the poorest continent, Africa, which already has water shortage and ecological damage.[10]

"HOT AND SICK": HEALTH AND GLOBAL WARMING

We will look at just one of the many scenarios, the effects on human health, because this illustrates the intricate relationship between damage to human health and the environment. Public health experts recognize that global warming is already killing 150,000 persons each year due to heat stroke, salmonellosis and other food poisonings, and malnutrition due to crop losses.[11]

Climate change–related pathways to ill health are likely to arise both relatively directly (e.g., via heat stress and injury from storms) and through complex mechanisms that involve disturbance to ecological systems. As explained in chapter 4, ecosystems provide services essential to the survival of humanity. Many ecosystems are already stressed by factors such as pollution, land use changes due to increased population and economic activity, bioinvasion, and loss of resilience consequent to altered biodiversity. For example, climate change may add to existing soil degradation due to overcropping by altering the soil's microbial ecosystems. This will further decrease crop yields needed to feed a growing human population. Some ecosystems will find it difficult to adapt and will undergo significant or irreversible damage. These include coral reefs and atolls, mangroves, boreal and tropical forests, wetlands, and native grasslands. The end result of all these changes is the possible impairment of human health by reductions in nutrition, economic activity, and habitable locations and an increase in infectious diseases. Many of these scenarios are likely to impact human health. For example, damaged and reduced coral reefs and mangroves will harm important fish-breeding habitats. Many poor coastal and island populations are dependent on fishing for most of their dietary protein, and they are thus vulnerable to malnutrition. Reduction of tropical forests increases the vulnerability of populations mostly in the developing world to erosion and flooding. Climate change–related loss of wetland and damage to native pastures will also change ecological systems essential for agriculture and aquaculture.

Many infectious diseases, both vector borne and spread by microbial-contaminated food and water, are sensitive to changes in climatic conditions and have been predicted to increase in incidence and regional seasonality. (A vector is an organism that transmits the disease.) As the global temperature increases, many vectors and potentially the diseases they carry will extend their geographical range. Recent modeling studies have indicated that malaria and dengue fever, which are spread by mosquitoes and which impinge on 40 to 50 percent of the world's population, are likely to increase especially in populations in the developing world, vulnerable because of poverty and unprotected by good public health resources.

Large changes to global temperature, extreme weather events, and a changing distribution of precipitation could exacerbate conflict, war, and the dislocation of people to become environmental refugees, as communities compete for diminishing supplies of fresh water and arable land. The projected increases in global temperature mean an increased frequency of heat waves, the health effects of which may be exacerbated by urban air pollution and increased humidity. This would lead to increases in heat-related deaths and illnesses, especially in populations unaccustomed to heat waves and made vulnerable by age, preexisting illness, or poverty.

Increased weather-related disasters, flooding, storms, and droughts are likely to increase especially in the third world. Poor housing and infrastructure, inadequate organizational and relief capability, and epidemics of malaria, diarrhea, and respiratory infections are likely to increase morbidity and mortality. In some cases, starvation and malnutrition could follow. However, the developed countries will also suffer an increase in injury due to storm. Hurricane Katrina, which devastated the city of New Orleans, killed 1,000 people and left a repair bill of around $235 billion.[12] According to the U.S. National Oceanic and Atmospheric Administration, 2005 was the most turbulent year on record. There were 26 named storms, 13 of which were ranked as hurricanes and 7 were strong enough to be ranked as major hurricanes.[13] Meteorologists are still divided over the question of whether this upsurge of hurricanes can be explained on the basis of global warming.[14]

THE DAY AFTER TOMORROW?

So far, we have discussed events that are predicted with a high degree of certainty. However science is discovering mechanisms that may result in sudden irreversible changes in the earth's environment. These are termed *threshold events,* whereby a further small increase in temperature triggers a major change in the earth's control mechanisms. The U.S. National Academy of Sciences supports this concept and believes that there could be an abrupt climate change.[15] The following are a few examples of possible mechanisms. The gulf stream, flowing north in the North Atlantic Ocean, warms northern Europe and returns deep cold waters flowing south. Studies from the National Oceanographic Centre in the UK have shown that the returning current may have slowed by 30 percent since 1957.[16] The northward flow is weakening due to climate-related increases in the southward flow of fresh water from melting ice. This event is depicted in the doomsday thriller *The Day After Tomorrow.* If the gulf stream reversed, Europe would have the climate of Hudson Bay, despite a warming world.

There are a number of natural stores of greenhouse gases ("sinks") in the tundra, soils, and oceans. These sinks could release their gases as the temperature increases, leading to a rapidly accelerating global warming. The permafrost in the tundra of Siberia is thawing rapidly and is releasing frozen stores of the greenhouse gas methane.[17] The oceans absorb 2 billion tons more carbon dioxide than they release each year, and this is about one third of all carbon dioxide produced by humanity. In future, with warming of the ocean's water, this sink may be compromised and there may be a net release of carbon dioxide into the atmosphere. However at present the Antarctic Ocean is becoming more acidic due to absorption of carbon dioxide from the atmosphere. The acidity will affect the ability of tiny crustaceans to grow their calcium carbonate shells, and an important link in the food chain may be lost.[18] The forests of the world are an important carbon sink, but as the temperature rises trees become sick and become net producers rather than storers of carbon dioxide. British scientists have also discovered another feedback mechanism whereby warmer temperatures have increased microbial activity in the soil, releasing greater than expected amounts of carbon—quantities sufficient to reduce Britain's attempt to curtail greenhouse gas emissions.[19]

There are other mechanisms whereby global warming is being accelerated. Arctic ice is rapidly melting, being 20 percent less than normal during the summer of 2005. Dr. Mark Serreze of Colorado's National Snow and Ice Data Center, believes that a threshold may soon be reached beyond which sea ice will not recover. A feedback process may be set in motion, accelerating the melting of ice, as there is more open blue water to absorb solar energy and less white ice to reflect sunlight back into space.[20]

The major threat of global sealevels rising comes from the glaciers of Greenland and Antarctica. Greenland's glaciers are melting into the sea at almost twice their previously observed rate in the last five years.[21] The average temperature of Greenland has risen by 3°C (5.4°F) over the last two decades and between 1996 and 2006 the amount of water lost from Greenland's ice sheet increased from 90 cubic kilometers (21.6 cubic miles) to 220 cubic kilometers (52.8 cubic miles) per year. Greenland's ice sheet covers

1.7 million square kilometers (0.66 million square miles) with ice of up to 3 kilometers thick (1.86 miles), and if completely melted it would raise global sea levels by around 7 meters (7.65 yards).

The evidence that we are moving into an accelerated phase of global warming is supported by data showing that 9 of the 10 warmest years since 1860 have occurred since 1990 and 19 since 1981, and annual increases in the concentration of carbon dioxide in the atmosphere are accelerating as shown by data from the U.S. Government's National Oceanic and Atmospheric Administration. These measurements are sufficient for scientists to be increasingly concerned that damage to carbon sinks and other mechanisms described above may be playing a part.

James Lovelock is a scientist, respected internationally for his pioneering work on biological feedback systems. He introduced the Gaia concept of the living earth acting like a single organism by using feedback mechanisms to maintain stability of temperature and climate over long periods of time. In 2006, in his book *The Revenge of Gaia,* he argued that global warming will be amplified by the simultaneous malfunction of several feedback systems due to human activities and it is already too late to stop catastrophic warming.[22] One such mechanism is that of global dimming, whereby aerosols in the atmosphere produced by global industry are shielding the earth from part of the sun's radiation. With a severe industrial downturn, a sudden leap in global temperatures will be expected. Various events are likely to precipitate economic downturn, such as the likely oil shortage are discussed later in this chapter.

Regardless of whether climate change is the serious problem accepted by most national governments or whether we are moving toward a catastrophic change as predicted by Lovelock will continue to be debated, there is little action to prevent it. Why not? As discussed in chapter 1 there are a number of psychological factors such as denial that prevent individual responses to potentially catastrophic events. However these responses do not account for the actions of world leaders. As researched by Beder,[23] prior to the Kyoto conference in 1997, a U.S. consortium of 20 fossil fuel organizations launched a campaign opposing the treaty on the basis that jobs would be lost and energy prices would rise. Thereafter corporations used front groups, public relations firms, and conservative think tanks to cast doubt on the science and impacts of global warming. The names of the organizations were Orwellian, "Advancement of Sound Science Coalition," "The Coalition for Vehicle Choice," "Global Climate Information Project," "The Greening Earth Society." The latter has stated that "using fossil fuels to enable our economic activity is as natural as breathing."[24] Senator James Inhofe, a conservative Republican, called human-caused global warming "a hoax."

He received an environmental award for his support of "rational, science-based thinking and policy-making" from the Annapolis Center for Science-Based Public Policy that receives funding from Exxon Mobil. Inhofe is chair of the Senate Environment and Public Works Committee.[25]

As with any scientific consensus, there will be dissidents. It would be expected that scientific conclusions that are in effect computer forecasts based upon existing data might be open to differing interpretations. Indeed detailed scholarly critiques of the conclusions have been published.[26] But the skeptics are a diminishing breed in the face of the mounting evidence from many scientists in many disciplines, and their task is difficult because in the industry campaign to derail Kyoto many, but not all, were well paid to travel the world to muddy the water by plying their wares in the media. Since the media sometimes try to operate on the basis of balance, they use opposing opinion even when there is only one opposing opinion to the views of a thousand scientists. This has often allowed skeptics to have more exposure to the public than their views deserved. Corporate think tanks such as the Heritage Foundation published in 1997, "The Road to Kyoto; How the Global Climate Treaty Fosters Economic Impoverishment and Endangers US Security."[27] The foundation predicted that Kyoto would cost as much as $30,000 in lost income per family per year. The Competitive Enterprise Institute wrote that "the likeliest global climate change is the creation of a milder, greener, more prosperous world."[28]

This was the background to George W. Bush's succession to office in early 2001. He was an oil man who appointed oil men to his cabinet and was heavily indebted to them for political donations. In the words of the late Robin Cook, former UK foreign secretary, "there has never been an administration with hands so dipped in Texas oil. There was a super-tanker somewhere out on the seven seas called the Condoleezza Rice."[29] The name of this Chevron tanker was changed to "Altair Voyager" when Ms. Rice was appointed national security advisor in 2001.

It was not surprising that the president's top policy was to increase the flow of petroleum from foreign suppliers to the U.S. market.[30] Bush established the National Energy Policy Development Group (NEPDG) chaired by Vice President Dick Cheney, formerly chair and CEO of Halliburton Oil. But even before the report, Bush questioned the scientific evidence of warming and said that Kyoto was unfair and too expensive for the U.S. economy. In 2001 he responded to a memorandum from Exxon asking that Dr. Robert Watson, chair of the IPCC, be replaced, because of his opinion that greenhouse emissions must be reduced.[31] Watson was replaced. The NEPDG did not propose any reduction in oil consumption. Instead it proposed to slow the growth in U.S. dependence on imported oil by increasing production at home by exploiting untapped reserves in wilderness areas. In effect Bush made the decision to increase his dependence on oil. This decision and the continuing corporate opposition to greenhouse reduction has dictated the government's decisions to oppose any climate change negotiations culminating four years later in the continued obstructionism to future negotiations at the climate meeting in Argentina in December 2004 and at the Montreal meeting of Kyoto parties in 2005. The Montreal meeting of 180 countries was intended to commence a new negotiation on greenhouse emissions to be implemented in 2012 when the Kyoto agreement terminates. The succeeding meeting in Nairobi in November 2006 also failed to draw a timetable for cuts in emissions. It is clear that the failure of the United States to participate and its lack of leadership is a major impediment to progress.

It would be wrong to conclude that the fossil fuel industries have influenced only U.S. policy and not that of other countries. European countries have signed Kyoto and have developed alternative energy programs, but one has to look to Australia, the other nonsigner of Kyoto, to see the malign influence. There the government relied heavily on figures and advice from the Bureau of Agricultural and Resource Economics (ABARE), funded by business and fossil fuel industries.[32] Places on the steering committee were offered for $50,000 each, and those who took advantage included Mobil, Exxon, Texaco, BHP, and the Australian Aluminum Council. As happened in the United States, ABARE predicted a huge loss of jobs and income if emission-reduction targets were to be met. The Australian government has worked secretly with the fossil fuel industry to produce an energy plan that will rely on geosequestration of carbon dioxide, with neglect of alternative energy.[33] Despite the strengthening of scientific evidence that human influence is causing global warming, determined resistance to these findings continues in the form of so-called scientific societies such as the George C. Marshall Institute in the United States and the Scientific Alliance in the UK.

In 2005, the United States and Australia, the two main antagonists of the Kyoto agreement, joined with China, Japan, India, and South Korea to form the Asia-Pacific Partnership for Clean Development and Climate. This rejects mandatory targets on greenhouse gas emissions and promotes technological solutions instead. Opponents of the partnership accept that technological solutions must be sought but see dangers in relying solely on such developments. At the first meeting of the partnership in Sydney in January 2006, India's environment minister announced that India will not implement mandatory emissions reduction of greenhouse gases. Since India is a signatory to the Kyoto agreement and is likely to have to

Worldwide discovery of oil peaked in 1964. The consensus opinion of ɪny oil scientists is that fewer major discoveries and increasing difficulties extraction of oil from existing fields will lead to decreased production in ᵻ next 5 to 15 years. The peak of a bell-shaped curve of production de-ibed by M. King Hubbart,[37] respected geologist and former oil industry ʊert, has been reached, and we will soon be on the down curve. The price oil will rise, for demand will increase with the burgeoning development China and India.

There is an alternative viewpoint that no crisis is imminent, that explo-ɪon and discovery on a needs basis continues, and that there are many ᵻre regions to be explored. Analysis of the situation is made difficult for ɪtical and commercial reasons. Many countries have inflated their re-ʋes because it allows them to obtain loans and to export more oil. Major npanies have also inflated their reserves to increase their share price and ʋernments have misused statistics to suit their purposes. However many ent reviews continue to support Heinberg's and Campbell's analyses.[38]

If indeed civilization is moving into an era of oil scarcity—and having ᵻsidered all the evidence the authors agree with this scenario—we have recognize that the present world population cannot be supported and need to reduce it could cause chaos and conflict. Heinberg ventures the ɪnion that a world population of only two billion would be sustainable.[39] ᴑy is this? Principally because oil has been the major factor in the dramatic ɪrease in food production that has facilitated and fed the world's growing ᴑulation. Oil provides the nitrogen fertilizer, the herbicides and pesticides, ᵻ fuel for tractors to service larger farms and for transport and distribution ᴑroducts. Humans, like other animals, expand their numbers according to abundance of nutrition available, and it can be postulated that the use of ɪil fuels and oil in particular has created the population explosion.

ᴑil then has been the linchpin for the expansion and technological de-ᴑpment of Western civilizations, and America's head start in discovering ᵻ utilizing its own oil resulted in its untold affluence and world domina-ᵻ. More than any other country, the United States is dependent on oil for ɪsport, personal and communal, electricity, and industrial and armament ᵻduction. A standard of living, so high that it would be incomprehensible ᴑ generations ago, has been achieved by the exploitation of this resource. to maintain its standard of living, the United States presently imports 53 ᵻcent of its oil needs and by 2020 this will have increased to 72 percent,[40] ᵻecarious position that has come to dominate U.S. foreign policy.

ᴴowever, oil production and consumption have resulted in damage to environment to such an extent that human civilization is under threat. ᵻ possession of oil resources potentially confers such power and wealth

adhere to mandatory reductions after 2012, the partne
as a mechanism to destabilize Kyoto and continue with
usual. This interpretation tends to be confirmed by the
ment over five years of $100 million by Australia and
United States to technological solutions compared to
lions invested in the war on terror.

There are many other factors operating in the Un
allowed this misguided policy to progress without sig
These will be analyzed in later chapters, but they are c
power, wealth, and influence of the fossil fuel indust
of Western civilization. We have chosen to analyze th
addiction to it, like all addictions, overwhelms rationa
the points we make are equally relevant to the coal ind

BLACK GOLD OR THE DEVIL'S EXC

The gold rushes of yesteryear are now superseded b
gold. The public image is one of the adventure of exp
in a huge explosion as the gusher blackens the sky wi
riches abound. In reality the image should be one of t
which determines the foreign policy, overt and covert
powerful nation. The need for oil is such that powerf
any length to get it no matter whom or what gets de
nomic power for it is the dynamo for economic grow
and the freedoms of neoliberalism have created a phil
dorial destruction of the environment, encapsulated ii
Vice President Cheney that it is the God-given righ
to consume as much cheap gas as they can while driv
they can find.[34] This attitude determines the calamities
more will follow because during this century the cor
change will occur together with those of the expecte
struggle to control and adapt to climate change will
with the convulsions caused by declining agriculture a
needed to support the world's population. The pressui
more damaging fossil fuel, coal, will be enormous.

Oil is a finite recourse and while there is debate a
out, the evidence suggests that it is likely to be soon. I
Richard Heinberg[35] and *The Essence of Oil and Gas De*
bell[36] it is argued that oil extraction will peak early t
by 2021. We will not know exactly when at the tim
in retrospect.

to governments and corporations that a mentality of denial has pervaded the consequences. Greenhouse emissions are seen as a necessary consequence just like the pollution of land, sea, and rivers. We will discuss the following indictments.

The exploitation in many poor countries has hindered their development and left a legacy of environmental degradation. In 1995 and 2001, Sachs and Warner studied data from 97 countries to measure their economic performance against their degree of dependence on the export of natural resources.[41] They showed that exploitation of natural resources suffocates other economic activity. The transport of oil through the shipping lanes has resulted in many disastrous spills with severe consequences. There are many examples of decades of pollution of land and coastal waters that have damaged that health and livelihood of populations. And finally, the unthinking pursuit of oil supplies has become an overriding factor in the foreign policy of the United States, with increasing consequences for all humanity.

OIL AND DEVELOPING COUNTRIES

In the report *Fuelling Poverty—Oil, Water and Corruption*[42] Christian Aid draws on data from the International Monetary Fund (IMF), World Bank, and its own experts to analyze the impact of oil discovery. In developing countries oil becomes a key ingredient in creating greater poverty for the majority of the population. There is increased corruption, a greater likelihood of war or civil strife, and a perpetuation of dictatorial or unrepresentative government. While the exploration and delivery of oil to developed countries has been carried out by multinational oil companies, such is the need for their success that $20 billion of public money has supported their enterprises. The Christian Aid report analyses oil discoveries in three countries, Angola, Sudan, and Kazakhstan.

In Angola, oil produces revenues of $5 billion per annum. but one billion of this goes missing. This money provides 90 percent of government revenue, and almost all consumer goods and services are imported indicating little development of local employment or infrastructure. The income from oil fueled a 30-year war. Oil spills have done environmental damage, and operating companies are not subject to environmental surveillance. Angola ranks as one of the world's poorest countries, and two thirds of the population do not have safe drinking water.

In Sudan, civil war was perpetuated and financed by oil, but exploitation continued amidst the carnage. In Kazakhstan there is abject poverty despite massive revenues. While continuing poverty has been a feature of oil discovery in developing countries, discovery of oil in small Western countries, such as Norway and Scotland (the Shetland islands) has lead to the creation of public trust funds to maintain community support and development once the wells were exhausted. In analyzing further the disastrous situation in developing countries where oil has been discovered, we must examine the role of the liberal democracies and their responsibilities in regulating the corporate empires that they support.

In the case of Angola, corruption is encouraged by signature payment to the government to win commercial advantages that are subject to confidentiality agreements. In all these matters the needs of rich countries and oil companies have overridden the need for reform. In an attempt to expose what is paid to governments, the UK government introduced an Extractive Industries Transparency Initiative. British Petroleum published details of its payments to Angola and was threatened with termination of contracts. The initiative has failed because the needs of companies for confidentiality in relation to competition outweighs their responsibility to individuals outside their country of regulation. It has to be asked why international finance and trade can be protected legally (because it suits the liberal democracies) when regulations on behavior cannot be imposed by Western governments, company law, and stock exchanges (because it wouldn't suit the liberal democracies). In its report, Christian Aid called for reforms such as the cessation of provision of public and IMF money to oil projects and for the creation of trust funds paid by the companies for the purpose of infrastructure development. UK Prime Minister Blair demanded voluntary disclosure of payments to repressive regimes, but in 2004 huge payments were still being made.[43] It is surely an indictment of liberal democracies that this situation has existed for decades without any significant attempt at reform by regulation.

Many countries other than Sudan and Angola have suffered from the eviction of indigenous peoples and conflict as a result of oil discovery, the need to protect pipelines, and the provision of arms to suppress opposition. Those involving the United States are Guatemala, Columbia, Democratic Republic of Congo, Aceh in Indonesia, Afghanistan, and Iran. France, Russia, and Britain have responsibilities for disturbance in other countries.[44] China has quickly developed oil interests in Angola, Sudan, Nigeria, and Algeria to feed its rapidly expanding economy[45] and seems destined to emulate the poor practices of the Western nations.

OIL POLLUTION: OIL AND NATURE DON'T MIX

The pollution of the seas, rivers, and wetlands during the exploration and transport of oil illustrates the disregard for the environment by oil companies and governments, mainly the liberal democracies that fail to control them, and it reflects an absolute priority to obtain this vital source of energy at whatever cost. We will analyze the cause of some of these events.

In an analysis "Lessons Not Learned: The Other Shell Report" by Friends of the Earth regarding the activities of Shell in nine countries, a conclusion is reached that "Shell continues to hold onto an industrial infrastructure that is hazardous to people and to the environment, to operate aging oil refineries that emit carcinogenic chemicals and other harmful toxins into neighborhoods, to neglect contamination that poisons the environment and damages human health, to endanger the survival of species, and to negotiate with local governments for substandard environmental controls."[46]

In Nigeria, it has been alleged that for decades the activities of Shell have contaminated land, forests, lakes, and mangroves around refineries and pipelines.[47] On November 14, 2005, the federal court of Nigeria ordered a group of oil companies including Shell to stop gas flaring in the Niger delta. Plaintiffs allege that this practice has been continued for decades causing huge greenhouse gas production and numerous deaths from local pollution. Plaintiff estimates for one small area of the delta suggest 49 premature deaths and nearly 5,000 respiratory illnesses per annum. It has been claimed that Shell has known for decades that the practice of flaring was harmful, yet has continued despite its statements on environmental responsibility. Shell's response to "The Other Shell Report" was that "we reject any links to human rights abuses in Nigeria or elsewhere."[48] In February 2006 a Nigerian Court ordered Royal Dutch Shell to pay $1.5 billion in damages for polluting the Niger delta.[49] In its Sustainability Report, 2005, Shell states, "We are committed to doing our part by funding environmentally and socially responsible ways to meet the future world's energy needs."[50] Shell has stated that it will cease flaring by 2008. Furthermore, on May 27, 2005, Shell joined with 12 other of the UK's most powerful companies in writing to Prime Minister Blair asking him to take urgent action on climate change.[51]

On March 24, 1989, the oil tanker *Exxon Valdez* ran aground in King William Sound, Alaska. As a result, 258,000 barrels of crude oil were spilled and rapidly dispersed by wind and currents throughout a region of commercial fishing, tourism, and biological treasures. Some 4,000 miles of shoreline were contaminated. Hundreds of thousands of marine birds were killed, together with thousands of sea otters, whales, and large fish. This spill occurred in a sensitive and economically important region of a liberal democracy, and its biological effects have been able to be carefully studied. After 15 years, there is only partial recovery.

It is important to recognize that the volume of this spill was only a small proportion of the spills occurring on land and in sea around the world each year. In the year following the *Exxon Valdez* spill, there were 10,000 smaller spills. All these exert continuous toxic pressure on oceanic and coastal ecosystems.[52] The record of oil spills in developing countries is appalling, and when environmental regulation is absent, little is done to mitigate them.

The large oil spills at sea result from the holing of tankers in collisions, tankers hitting rocks, or ships going aground. The *Exxon Valdez* under the command of a drunken captain ran onto rocks at full speed and would have been holed even if it had a double hull. But it is recognized that in most instances a double hull prevents leakage. Since the *Exxon Valdez* disaster most major marine oil spills have been from aged single-hulled vessels.

The *Sea Empress* in 1996[53] and the *Prestige* in 2002[54] illustrate the factors involved. Both were single-hulled ships. The *Sea Empress* struck a charted rock in Milford Haven Harbour, Wales, and 70,000 tons of light crude oil was spilled resulting in extensive coastal pollution. The vessel was Spanish built, Norwegian owned, registered in Cyprus, French chartered, managed in Glasgow, Liberian flagged, and had a Russian crew. There was a communication problem between the Russian-speaking captain and the pilot. The *Prestige,* a 26-year-old vessel split and sank off the Atlantic coast of Spain with the release of 60,000 tons of heavy crude oil. There was extensive pollution of the coasts of Portugal, Spain, and France. The *Prestige* was crewed by Filipinos, the captain was Greek, and ownership was Liberian but controlled by a Greek cartel and registered under a Bahamian flag. Shipping companies use such flags of convenience for registration in tax-free havens and to avoid maintenance of old vessels and regulations that govern the training and conditions for crews.

Pollution by developing countries is a sad indictment of the ethics and greed of Western corporate empires. When pollution occurs in Western countries there is some degree of accountability through the justice system, though it must be recognized that wealthy companies use legal options to delay judgment and minimize compensation. The *Exxon Valdez* case is still in the courts. In other instances there is no agreement on the remedy, for example millions of tons of oil waste lie on the bed of the North Sea, a legacy of Britain's oil boom.[55] The oil and heavy metals have severely damaged large areas of the seabed and are partly responsible for the decline of fish stocks. Yet many such events occur in many parts of the world with little attempt at assessment or remedy. It is a reflection of the power of the oil industry that little attempt has been made to impose environmental accountability on these companies. The liberal democracies in which the industries reside have not recognized their responsibilities.

Regulation of aging and potentially unsafe oil tankers must be discussed in terms of the record of democracies in protecting the environment.

International safety regulations have been vested in the International Maritime Organization (IMO) a UN body.[56] Under IMO regulations, national port inspectors can check for safety hazards and impound the ship. In the case of the *Prestige*, British Petroleum inspected it in 2000 and found it totally unacceptable for its own use. It was inspected by the American Bureau of Shipping a year before it sank,[57] and it was only operating off the coast of Europe under proposals that were watered down after intense lobbying from shipping and financial interests.

The crux of the problem is identified by Caroline Lucas, Member of the European Parliament, "Really, we are all to blame for allowing our (usually elected) politicians to put the short term priorities of businesses and financial markets before the long term economic, social and environmental needs of us all. It is the globalization of this trend over the past decade or so that I hold accountable for the fate of the Prestige."[58] David Cockroft, general secretary of the International Transport Workers Federation agrees when he says, "Shipping was the world's first globalized industry... But it is characterized by a culture of secrecy and deceit... Shipping's culture of secrecy means that when criminally dangerous ships sink, owners can hide behind the brass plates that make each ship a separate company, company owners can then declare bankrupt... Then the owners can carry on trading with their other ships."[59]

Oil pollution is a solvable problem. It remains unsolved because financial interests pressed by powerful voices override long-term environmental considerations. The IMO does not have the unconditional support of the United States, and nations do not fulfill the obligations of the IMO. Are we saying that the regulations cannot be rigorously enforced on every ship visiting American and European democracies? Or do we prefer not to?

The issue of double hulls in oil tankers is one in which environmental groups have struggled against inertia. In 2003 an agreement entered into force for 130 participating countries to phase out single-hulled tankers of over 200,000 tons that were at least 23 years old. In 2005, these regulations were tightened.[60] Whether this will reduce pollution remains to be seen, for legal agreements that would enforce oil tankers to have double hulls have essentially been avoided through the use of "flags of convenience." Through a complex maze of companies and subsidiaries, ownership of tankers can be hidden. It is not known whether a particular ship has a double hull until it is too late and its single hull is sliced open in an accident that pollutes hundreds of square miles of oceans and coastline. Then lawyers cannot find the owners and legal redress becomes impossible.

This evasion is possible because of the structure of modern corporate law. From about 150 years ago, courts in Britain decided that a corporation was legally distinct from its owners, directors, and members. This created the notion of a distinct legal identity for corporations. Corporate law was drafted by lawyers who were in the pockets of big business. This led to a situation where corporations became autonomous, like robots in movies that suddenly become alive. The structure of corporate law of most Western nations is more complex than any other body of law (with the possible exception of taxation), and it is relatively easy for corporate lawyers to hide the ownership of oil tankers and indeed any other investment. All of this is legal and is permitted by corporation law.

Liberal democracy is implicated in this process of ecological deceit once more because of its intimate association with global capitalism. Liberal democratic beliefs influenced those drafting corporation law to allow the free movement of property and goods across the world. And of course liberal democratic ideology influenced those representative members of our governments who voted for such laws. It is this legal framework, especially designed to allow corporate power to flourish, which permits the flags of convenience and their owners to escape responsibility.

BUSINESS AS USUAL: THE VENEER OF CARE

Western democracies have failed in their responsibilities to safeguard the common good. As in many spheres of finance and trade, Western democracies practiced some social justice regulation and democratic values at home but have not applied these tenets of their democracy to developing countries. This is a recurrent theme in *The Roaring Nineties* by Joseph Stiglitz.[61] It is argued by Western democracies that it is inappropriate to interfere with the ways sovereign countries conduct their affairs and do business. We disagree and maintain that it is a failure of liberal democracy. The story is familiar. We are addicted to oil. Therefore those who produce it are powerful. They add to this power with huge financial support to political parties.

It is relevant to ask whether democracies have made any effort to substantially reorient their policies to energy efficiency and renewable energy sources. In the 1980s Denmark began to develop wind energy that now provides 10 percent of its electricity, and other European democracies have developed similar programs. Sweden is to take the biggest energy step of any advanced Western economy by trying to wean itself off oil completely within 15 years—without building a new generation of nuclear power stations. The intention is to replace all fossil fuels with renewable energy before climate change destroys economies and growing oil scarcity leads to huge new price rises.[62] However these efforts are a drop in the ocean of necessary reform.

In the United States, President Bush has said, "We need an energy policy that encourages consumption."[63] To ensure its supplies, the United States has progressively increased its military bases around the world and it has prosecuted war and disorder in other countries. It has illustrated that democracy provides no barrier to executive decisions that can threaten the stability and future of the world. The actions of a wealthy country that considers only its own needs and a president predominantly concerned more with maintaining his own power than with ensuring the future of the world will be condemned by all who understand the crisis.

But is easy to condemn. It does not take us further into understanding the underlying nature of the problem. There are those who suggest that mandatory corporate responsibility will be important in controlling destructive practices,[64] but there is no sign that governments will accept this. In the UK, the Blair government introduced a Company Law Reform Bill in response to pressure from the Corporate Responsibility Coalition of 130 UK civil society organizations, including Amnesty International UK, Friends of the Earth, Christian Aid, unions, and academic institutions. In this bill it was proposed that company directors be given a legal "Duty of Care" to communities and the environment as well as for victims of corporate wrongdoing to be given legal rights of redress. The original bill was diluted as a result of lobbying by some corporate groups so that directors were required by the bill only to "have regard to" communities and the environment. A modest reform became a sinecure.[65] In any event, a history of corporatism indicates that regulations are there to be demolished or evaded.[66]

We have to understand the extreme paradigm of belief that places greed and success before the rational need of others. We will give two brief examples, one from the United States and the other from Australia, that illustrate the collusion between corporate interests and democratic government to the detriment of democratic process and the interests of the public. An oil industry lobbyist, a lawyer with no scientific training, was transferred to the White House to become responsible for the Council on Environmental Quality. His task was to edit scientific papers to play down the dangers of global warming.[67] In Australia evidence has been presented that "for a decade the Howard Government's policies have been not so much influenced but actually written by a tiny cabal of powerful fossil fuel lobbyists representing the very corporations whose commercial interests would be affected by any move to reduce Australia's burgeoning greenhouse gas emissions."[68]

The extreme self-serving paradigm explains the pollutions, corruptions, and collusions described above. In 2006, Shell had a record profit of

$23 billion, yet its environmental record is seriously flawed, particularly in developing countries. George Monbiot has detailed how some oil companies have rebranded themselves by accepting the science of climate change and by issuing sustainability reports. However the environmental and human rights impacts of their operations remains essentially unchanged.[69] In general corporate leaders are able to subsume the health of populations and the future of their children to that of profit. It is a powerful belief system that drives these actions. Soon we will analyze further how liberal democracy has facilitated these events.

NOTES

1. J.T. Houghton, et al. (eds.), *Climate Change 2001: The Scientific Basis* (Cambridge University press, Cambridge, 2001), p. 90, available at <http://www.ipcc.ch>.

2. Ibid., p. 92.

3. International Society of Doctors for the Environment (ISDE), Position Paper on Climate Change and Human Health, 2002, at <www.dea.org.au>.

4. Intergovernmental Panel on Climate Change, "Climate Change 2007. Fourth Assessment Report," 2007, at <http://www.ipcc.ch/>.

5. B. McKibben, "The Coming Meltdown," *The New York Review of Books,* vol. 53, no.1, January 12, 2006, at <http://www.nybooks.com/articles/18616>.

6. International Society of Doctors for the Environment, Position Paper on Climate Change and Human Health, Policy and Actions, 2002, at <www.dea.org.au>.

7. Chapter 4 in J. Smith and D. Shearman, *Climate Change Litigation: Analysing the Law, Scientific Evidence and Impacts on the Environment, Health and Property* (Presidian Legal Publications, Adelaide, Australia, 2006).

8. Alan J. Thorpe, "Climate Change Prediction: A Challenging Scientific Problem," Institute of Physics, London, 2005, at <http://www.iop.org/activity/policy/Publications/file_4147.pdf>.

9. Ross Gelbspan, *Boiling Point* (Basic Books, New York, 2004).

10. "Climate Change Will Hit Africa Hardest," *Guardian Unlimited,* February 2, 2005.

11. D. Campbell-Lendrum, et al., "How Much Disease Could Climate Change Cause?" in A.J. McMichael, et al. (eds), *Climate Change and Health: Risks and Responses* (World Health Organization, Geneva, 2003), pp. 133–158.

12. S. Connor, "Scientists Warm to Hurricane Theory," *The Independent Weekly,* December 11–17, 2005, p. 10.

13. Ibid.

14. K. Emanuel, "Increasing Destructiveness of Tropical Cyclones Over the Past 30 Years," *Nature,* vol. 436, August 4, 2005, pp. 686–688; R.A. Pielke, "Meteorology: Are There Trends in Hurricane Destruction?" *Nature,* vol. 438, 2005, p. E11; C.W. Landsea, "Hurricanes and Global Warming," *Nature,* vol. 438, 2005, p. E11.

15. Committee on Abrupt Climate Change, National Research Council, *Abrupt Climate Change: Inevitable Surprises* (National Academies Press, Washington DC, 2002).

16. H.L. Bryden, et al., "Slowing of the Atlantic Meridional Overturning Circulation at 25 N," *Nature,* vol. 438, 2005, pp. 655–657.

17. F. Pearce, "Dark Future Looms for Arctic Tundra," *New Scientist,* January 21, 2006, p. 15.

18. C. Sabine, et al., "The Oceanic Sink for Anthropogenic CO_2," *Science,* vol. 305, 2004, pp. 367–371.

19. "Greenhouse Emissions Break Kyoto Vows," *New Scientist,* October 8, 2005, p. 7.

20. McKibben, "The Coming Meltdown," from note 5.

21. E. Rignot and P. Kanagaratnam, "Changes in the Velocity Structure of the Greenland Ice Sheet," *Science,* vol. 311, February 17, 2006, pp. 986–990.

22. James Lovelock, *The Revenge of Gaia* (Allen Lane, London, 2006).

23. S.D. Beder, "Corporate Highjacking of the Greenhouse Debate," *The Ecologist,* vol. 29, 1999, pp. 119–122.

24. Ibid.

25. "James Inhofe: Conservative GOP Senator Again Blasts Environmentalist Fearmongers," *Human Events Online,* January 5, 2005, at <http://www.humanevents.com/>.

26. Christopher Essex and Ross McKitrick, *Taken by Storm. The Troubled Science, Policy and Politics of Global Warming* (Key Porter Books Limited, Toronto, 2002).

27. Angela Antonelli, Brett D. Schaefer, and Alex Annett, *The Road to Kyoto: How the Global Climate Treaty Fosters Economic Impoverishment and Endangers US Security,* The Heritage Foundation, October 6, 1997, at <www.heritage.org/Research/PoliticalPhilosophy/BG1143.cfm>.

28. Ibid.

29. Robin Cook, "Special Relationship a Fantasy," *Guardian Weekly,* November 19, 2004, p. 13.

30. Michael Klare, "Bush-Cheney Energy Strategy: Procuring the Rest of the World's Oil," *Foreign Policy in Focus Magazine,* 2004, at <www.fpif.org>.

31. Paul Brown, "Oil Giant Bids to Oust Expert on Climate," *Guardian Weekly,* April 11, 2002.

32. Clive Hamilton, *Running from the Storm: The Development of Climate Change Policy in Australia* (UNSW Press, Sydney, 2001).

33. Andrew Fowler, "Leaked Documents Reveal Fossil Fuel Influence in White Paper," ABC Online, September 7, 2004, at <http://www.abc.net.au/pm/content/2004/s1194166.htm>.

34. William Rivers Pitt, "The Prophesy Of Oil," *Truthout Perspective,* March 7, 2005.

35. Richard Heinberg, *The Party's Over: Oil, War and the Fate of Industrial Societies* (New Society Publishers, Gabriola Island, BC, Canada, 2003).

36. C.J. Campbell, *The Essence of Oil and Gas Depletion* (Multi-Science Publishing, Essex, England, 2004); and C.J. Campbell, *Oil Crisis* (Multi-Science Publishing, Essex, England, 2005).

37. M.K. Hubbert, *Resources and Man* (National Academy of Sciences and National Research Council, 1969).

38. J. Leggett, *The Empty Tank* (Random House, New York, 2005).

39. Heinberg, *The Party's Over,* from note 35.

40. R. Freeman, "Will the End of Oil Mean the End of America?" March 1, 2004, at <http://www.commomdreams.org/cgi-bin/print.cgi?file = /views04/0301–12.htm>.

41. J. Sachs and A. Warner, *Natural Resource Abundance and Economic Growth* (Harvard University, Cambridge, MA, 1995).

42. "Fuelling Poverty—Oil, War and Corruption," Christian Aid Report, 2003, at <www.christian-aid.org.uk>;

43. "Time for Transparency: Coming Clean on Oil, Mining and Gas Revenues," *Global Witness Report 2004,* March 2004, at <http://www.globalwitness.org/media_library_detail.php/115/en/time_for_transparency>.

44. "Oil Wars," *The Ecologist,* April 2003.

45. Decian Walsh, "China's Scramble for African Oil," *Guardian Weekly,* November 18–24, 2005.

46. Friends of the Earth, "Lessons Not Learned: The Other Shell Report," 2004, at <http://www.foe.co.uk/resource/reports/lessons_not_learned.pdf>.

47. Friends of the Earth, "Lessons Not Learned: The Other Shell Report," 2004, referred to in "Oil Search Wrecking Nigeria," *Guardian Weekly,* June 18, 2004. The allegations referred to in the text are made by various groups and authorities cited in this article and at note 46.

48. See <http://www.euractiv.com/en/socialeurope/meps-campaign-step-corporate-respon sibility/article-141766>.

49. Rory Carroll, "Shell Told to Pay Nigerians $1.5bn Pollution Damages. Oil Giant Will Appeal Against Court Decision. Kidnap and Sabotage Cripple Production," *The Guardian,* February 25, 2006.

50. See <http://www.shell.com/static/envirosoc-en/downloads/sustainability_reports/shell_ report_2005.pdf p.2>.

51. Corporate Leaders Group on Climate Change," May 27, 2005, Prince of Wales's Environment Programme, at <https://www.cpi.cam.ac.uk/bep/downloads/bep_report_2005.pdf>.

52. P.R. Ehrlich and A.H. Ehrlich, *Healing the Planet: Strategies for Resolving the Environmental Crisis* (Addison-Wesley, Reading, MA, 1991).

53. TED Case Studies, Empress Case, at <http://www.american.edu/TED/walesoil.htm.>.

54. Caroline Lucas, "The Laissez-Faire Misadventure," Redpepper Archive, at <http://www.redpepper.org.uk/intarch/x-shipping-disasters.html>.

55. Fred Pearce, "Toxic Legacy of Britain's Oil Boom," *New Scientist,* December 7, 1996.

56. Jim Morris, "Ship Shape. UN Agency Takes on the Complex Task of Regulating the Far-Flung Shipping Industry," *Houston Chronicle,* 1996.

57. Eric Scigliano, "Puget Sound's Rustbuckets," *Seattle Weekly,* January 1, 2003, at <http://www.seattleweekly.com/features/0301/news-scigliano.shtml>.

58. Brown, "Oil Giant Bids to Oust Expert on Climate," from note 31.

59. David Cockroft, "More Than Just a Loss of Prestige," *Red Pepper,* January 2003, at <http://redpepper.org.uk/intarch/x-shipping-disasters.html#cockroft>.

60. "Global Phaseout of Older Single Hull Tankers Begins," *Environment News Service,* April 7, 2005, at <http://www.ens-newswire.com/ens/apr2005/2005-04-07-02.asp>.

61. Joseph Stiglitz, *The Roaring Nineties* (Penguin Books, London, 2003).

62. John Vidal, "Sweden Plans to be World's First Oil-Free Economy," *The Guardian,* February 8, 2006.

63. George W. Bush, speech in Trenton NJ, September 23, 2002.

64. "Oil Wars," from note 44.

65. "Company Law Reform Fails People and the Environment," Friends of the Earth, November 3, 2005, at <http://www.foe.co.uk/resource/press_releases/company_law_reform_fails_ p_03112005.html>.

66. J. Bakan, *The Corporation: The Pathological Pursuit of Power and Profit* (Constable and Robertson Ltd, London, 2004).

67. Julian Borger, "Oil's Spirit Burns on in the White House," *Guardian Weekly,* June 24–30, 2005.

68. Clive Hamilton, "The Dirty Politics of Climate Change," Speech to the Climate Change and Business Conference, Adelaide, Australia, February 20, 2006, at <www.tai.org.au>.

69. George Monbiot, "Oil Giants Spin Web of Deceit," *Guardian Weekly,* June 23–29, 2006.

— 3 —

Hunger and Thirst: The Theft of Food and Water

They hang the man and flog the woman
That steal the goose from off the common,
But let the greater villain loose
That steals the common from the goose.

—*English Folk Poem*, ca. 1764

THE EMPTY WELL

Clearly in 1764 there was a problem with the commons. This chapter will demonstrate that liberal democracy has not resolved the problem but worse, has aided and abetted the villain. We look at "God-given" sustenance from fish and water and determine how their inequitable use has come about.

The need for water is vital for every form of life on the planet. To remain alive, the human body has to lose water in urine and feces every day to rid itself of toxic wastes, and the water lost from the skin and lungs maintains the body at a constant temperature. If these losses are not continuously re-placed, changes in the salt concentration in the blood stimulate the brain to compel the individual to drink. Thirst is a compulsion far greater than that of hunger, for lack of drinking water threatens life within days, whereas the body can survive without food for weeks. Humans experience thirst as a compelling desire that cannot be resisted, so much so that mariners will drink seawater to their detriment, and those marooned in the desert will drink urine. Millions of the world's poor will drink filthy water laden with pathogenic bacteria because they have no choice. This adds to the burden of infection, ill health, and death in poor communities.

The Universal Declaration of Human Rights delivered in 1948 addressed every conceivable right of importance to those who wrote it—property, government, marriage, legal rights, and so on—but access to the commons of clean air and safe drinking water is not mentioned. Article 25 states that "everyone has the right to a standard of living adequate for the health and well-being of himself and of his family, including food, clothing, housing and medical care and necessary social services."[1] Presumably the right to safe drinking water is subsumed under the mention of food but the wording of the document identifies it as a litany of democratic needs of the individual. Fifty years later, in 2003 at the World Water Forum, Mikhail Gorbachev urged enshrinement of the right to water in the Universal Declaration.

Access to water supplies including safe drinking water is surely the most basic of human rights, yet 1.1 billion of the world's 6 billion people lack safe drinking water.[2] This problem goes hand in hand with a lack of sanitation for 2.4 billion people and underlies some 600 water-related deaths each day, mostly in children under the age of five. A child born in the developed world consumes 30 to 50 times the water resources of one in the developing world, and in many regions water quality continues to deteriorate. Yet the provision of water resources is central to poverty reduction and must be achieved before other advances are made. There are estimates that safe drinking water and sanitation could be provided in accordance with the Millennium Goals for about $7 billion per year[3] compared to the $843 billion yearly world expenditure on arms.[4] It is pertinent to analyze how the liberal democracies have and can address this issue.

Firstly, it is important to understand that water is a renewable resource because rainfall is recurrent. Falling rain is absorbed into the subsoil to replenish the water table and can then sustain plants and trees till the next rainfall. Intensive use of water has caused the water table to fall in every continent with effects on the productivity of food (and on some cities that are steadily sinking due to drying of the subsoil). As a consequence, the ground water has been further exploited by pumping to depletion in many areas of India, China, and the Middle East. Rain also soaks into the sponges of wetlands and forest floors to be released slowly into streams, rivers, and lakes; some is collected in underground stores (aquifers); all these sources are used for human and agricultural use. Currently humans use half of all runoff water but much of the other half is in inaccessible regions where it is not needed, such as Alaska, Greenland, the Canadian Arctic, French Guinea, Iceland, Guyana, Surinam, and the Congo. Currently available runoff water is about

6,900 cubic meters (7,546 cubic yards) per person per year, which would be just enough for everybody if water and people were evenly distributed. Taking into account the growth of the population and the need to provide water to grow food, in 25 years there will be 3 billion people living in water-stressed countries.[5]

Although rainfall will continue to be a renewable resource, though more variable in its distribution due to global warming, there are numerous human activities that are compromising the efficient use of this water. Degradation of soils due to overplanting of crops, clearing of forests, and draining of wetlands destroys the sponge so that water runs away quickly and causes flooding with further damage to soil and crops. The number of significant flooding episodes has increased worldwide each decade from 6 in the 1950s to 26 in the 1990s. These floods cause huge economic loss to developing countries, loss of life, and pollution of existing water supplies. Climate change is adding to the problem with, on the one hand, heavier monsoon rain and storm surges, and, on the other hand, an increase in droughts.

In analyzing the availability of runoff water, it must be recognized that this water is not just for human use. It is needed by the earth's ecosystems that support human life. The use of river water for human and agricultural consumption is causing many rivers to die; that is, their ecological systems are compromised to such an extent that their fish and the plants that support their banks do not survive. Furthermore the reduced flow to the sea compromises the health of the shorelines and the breeding grounds of fish. This problem is created because a balanced human diet of 3,000 kcal per day requires 3,500 liters of water per day for its production, which is 70 times more that the 5 liters required for other daily needs. This demand is putting increasing pressure on water needed for ecological services (see chapter 4).

Humanity has bolstered its supplies of water by building dams. Throughout the world there are now 45,000 large dams defined as more than 15 meters (16.4 yards) in height and 800,000 small dams.[6] These provide water for drinking, irrigation, industrialization, and hydropower, and they mitigate flooding. But there are negative consequences, immediate and long term. Building dams has resulted in extensive loss of wetlands and displacement of entire communities and loss of their productive land. In many instances, the natural flow in rivers has been reduced to a trickle, and fish, an important food source, have disappeared. The dams on the Nile are changing for worse the productivity of the soils of Egypt and particularly the delta because the land requires deposition of silt from floods to maintain its fertility. The lack of river flow into the eastern Mediterranean Sea has damaged the marine ecology. Many examples are emerging of salination from the use of dam water over many decades and many soils are being rendered useless. Some nations are using dams to capture water from their neighbors. In the case of the Mekong River, giant Chinese dams are threatening the livelihoods of 100 million people in southeast Asia.[7] Flow through Burma, Thailand, Laos, Cambodia, and Vietnam is reduced, and ecological parameters are deteriorating. The Mekong is crucial in providing fish protein to these countries, and the seasonal flooding is necessary for the cultivation of rice.

There is one further source of safe drinking water. There are vast underground aquifers of water. In most instances the water was deposited in these fossil aquifers in previous geological eras, and it is replenished very slowly or not at all. Their average recharge time is probably around 1,400 years.[8] Ground water is used to supplement drinking water for irrigation and for industry. Many countries depend on groundwater for irrigation, and in India it is used to irrigate half of all arable land. Irrigation on the Great Plains of North America has relied on drilling into the Ogallala Aquifer, the largest in the world, which is now depleting,[9] as are groundwater levels in China, North Africa, and the Middle East. Many large cities including Beijing, Mexico City, Manila, Bangkok, Seoul, and Jakarta are threatened by reduced water tables.

We will now analyze the role of liberal democracy in assuaging this water crisis. Water is part of the global commons, and a full understanding of this concept will be provided in chapter 5. But meanwhile we will analyze the use of groundwater, the capture of water into dams, and the distribution and sale of water.

MINING FOR WATER

Aquifers are being used with little scientific assessment as to their natural renewal and therefore lifespan, in both liberal democracies and nondemocratic systems.

The fate of the Ogallala Aquifer in the United States depicts the attitude to an available but precious resource in the world's premier advocate of democracy. Its water is mined for profit like oil without consideration for the needs of future generations. As climate change brings a decreased rainfall in some regions, democratic decisions to use these resources are made ad hoc without long-term planning or sustainability requirements. In the southwest of Western Australia, aquifers are increasingly used for agricultural purposes without adequate estimates of their renewal rates. Australia, the driest continent, has several large aquifers. In South Australia, a dry state that depends partly upon the viability of a sick river, the Murray, for its water supply, the Roxby mine uses 30 million liters of water per day from the Great Artesian Basin. This source is a closed system created in geological time.

The water exists in fissures in the rock that close when the water is withdrawn and are then unavailable for replenishment.[10] Nevertheless, water supply to the mine was guaranteed by means of an Indenture Act that in effect overruled environmental considerations. A proposed expansion of the Roxby mine has been announced. This will require a further 70,000 million liters (7.9 million U.S. gallons) per day. The aquifer will be spared further depletion because the water requirements of the mine are likely to be provided by desalination plants. The energy requirements for these will increase greenhouse emissions. Prosperity today in prosperous countries has primacy over sustainability.

THE DAMNING OF DAMS

A large dam is 15 or more meters (16.4 yards) high or has a volume of more than 3 million cubic meters (3.28 million cubic yards). There are 45,000 large dams and they store 15 percent of the world's freshwater run off. The flow in 172 of the 292 largest rivers is regulated by dams.[11] There is economic logic for building dams to catch and use the runoff and flood water that otherwise goes to waste. This water can be harnessed for immediate economic gain through irrigation and energy generation. In 2000 the World Commission on Dams reviewed the effectiveness of the development of large dams.[12]

It was concluded that dams have made an important and significant contribution to human development, but in too many cases an unacceptable and unnecessary price has been paid to secure the benefits. The price was paid in both social and environmental terms. In one third of countries, hydropower provides more than half of the electricity supply and, overall, dams generate 19 percent of electricity, a contribution that needs to be maintained to reduce greenhouse emissions. Dams consume pasture, fertile land, and forests. They damage aquatic, terrestrial, and coastal species and so reduce biodiversity. They decimate river fisheries. These are not a significant problem in the headwaters in the barren glacial valleys of Switzerland or Canada, but in crowded developing countries the commission estimates that large dams have displaced between 40 and 80 million mostly poor people from subsistence living into abject urban poverty. Millions more living downstream have suffered serious harm to their livelihoods. The benefits to developing nations have been ephemeral. The dams have a finite life and as time passes they require increasingly expensive repair and are subject to sedimentation. The adverse impacts have been particularly devastating in Asia, Africa, and Latin America, where existing river systems supported local economies and the cultural way of life of a large population.[13] In these countries the poor vulnerable groups and future generations are likely to bear a disproportionate share of the social and environmental costs of large dam projects without gaining a share of economic benefits.

In the United States selective dams are being removed to improve the health of rivers and to allow free movement of salmon, an important food and recreational industry. By contrast, in India and China dam building continues undaunted by the loss of productive land. China has 46 percent (22,000) of the world's large dams and India 9 percent. The Report of the World Commission on Dams examined their effects on food, security, and nutrition. Between 1970 and 1995 nutritional status of the population increased by 14 percent in India and 30 percent in China, but it was not possible to attribute these improvements to irrigation agriculture from dam water.[14]

The recognition of these problems has not stopped the liberal democracies from financing large dams in developing countries through their own banks and the World Bank. The balance of scientific evidence is weighted more and more against such developments as their long-term environmental consequences become apparent and outweigh the benefits of electricity generation and industrial development. Furthermore the predicted technical, financial, and economic performance is below expectation. A consideration of all relevant factors, such as the need to maintain ecological systems, local food production, and social cohesion, as the bulwarks for the sustainability of society should lead to a re-evaluation of this form of aid to developing countries. Why then do these developments continue? The World Bank has to be seen as an instrument of the liberal democracies that benefits the construction companies of its powerful member countries. It is yet another example of the dichotomy of policies between home and abroad, driven by the common interests of corporate empires and governments. The performance of India, a liberal democracy that has displaced some 33 million people, is no better in social and environmental terms than that of totalitarian China. Certainly the Indian economy has progressed, helped by electricity generation, but whether this was an appropriate trade-off for the loss of productive land has to be doubted. Democratic institutions in the donor and the recipient countries have failed to provide sustainable outcomes.

The most important example that we can study is the damming of the Namada River in India described by Arundhati Roy in *The Algebra of Infinite Justice*.[15] Roy describes Indian democracy as the benevolent mask behind which a pestilence flourishes. The pestilence is the displacement of 33 million people by big dams in 50 years. Roy asks why damming of the Namada River by hundreds of additional dams proceeds when "big dams are obsolete...they're undemocratic. They're a government's way of accumulating authority (deciding who will get how much water and who will grow what where). They're a guaranteed way of taking a farmer's wisdom away from him. They're a brazen way of taking water, land and irrigation away from the poor and giving it to the rich."[16]

Roy points out that big dams account for only 12 percent of India's total food grain production. Yet the loss of other crops is huge due to the displacement of farming families. Because of widespread community opposition to one of the major dam projects, Sardar Sarovar, the World Bank instituted an independent review. Indeed the Bank had offered to loan money for the projects in 1985, two years before it was approved by the Indian Ministry of the Environment. The finding of the review in 1992 was as follows:

We think the Sardar Sarovar projects as they stand are flawed, that resettlement and rehabilitation of all those displaced by the Projects is not possible under prevailing circumstance, and that environmental impacts of the Projects have not been properly considered or adequately addressed. Moreover we believe that the Bank shares responsibility with the borrower for the situation that has developed...it seems clear engineering and economic imperatives have driven the project to the exclusion of human and environmental concerns...as a result we think that the wisest course would be for the bank to step back from the Projects and consider them afresh.[17]

The World Bank continued its support, though after a second report it withdrew in 1993, but funding continues from state governments.

How could such a project arise and be continued? Roy maintains that dam building in the Western world is in trouble, and so its beneficiaries have transported it to the developing world in the name of development aid. A nexus of politicians, bureaucrats, and dam construction companies all benefit by $20 billion per annum.[18] The World Bank provides the financing. Often other needs of the recipient country are tied to the deal, for example the provision of arms. Similar criticisms could no doubt be applied to many other dams constructed in the developing world, but there are few reviews like the one of Sardar, which was forced by vigorous community opposition. More frequently the displaced populations are suppressed into accepting their fate. The Three Gorges dam in China will displace 1.3 million people. The World Bank and a large number of Western democracies and financiers support the project.[19]

When dams are constructed on a river that runs through several countries it is not surprising to find potential conflicts. China, because of its might, surely recognizes that it can get away with damming the headwaters of the Mekong. Geoffrey Dabelko, Director of the Environmental Change and Security Project, the Woodrow Wilson International Center for Scholars, has stated that water is rarely the cause of major conflict between countries.[20] However, conflicts over water are expected to increase because of climate change and population pressures. Turkey plans to dam the Euphrates to the detriment of Syria, China and India have a potential conflict over the Brahmaputra, Ethiopia and Egypt over the Nile, and Angola and Namibia over the Okavango basin. The 5 percent of the world's population who live in the Middle East live on 1 percent of the world's water. Israel controls the river Jordon and during scarcity has cut supplies to Palestine and Jordon. Of 21 military events due to conflict over water, 17 have involved Israel.

WATER FOR SALE

The problem of water depletion is not solely a problem facing liberal democratic societies, as nondemocratic societies such as China and almost all Middle Eastern societies such as Saudi Arabia face the same intense problem. Water depletion is a direct product of an increased demand for water produced by the dictates of the economic necessity for increased economic growth. In the case of Africa the impetus is produced primarily by a need to meet the increased water needs of expanding human populations in a continent with poor resources. But in much of the world, the water problem is a product of the demands of national survival in the global economy. China's existing and future water problems, especially the poisoning of most of its major rivers[21] arises from the demands of producing consumer goods for affluent Western nations. Liberal democracy is implicated in the environmental destruction, because, as we have argued previously, it is a philosophical system that has rationalized and justified the globalist free market system that now pillages the planet. Behind the rhetoric of the global village lies the global pillage.

Liberal democracy as the theology of unrestrained global capitalism has failed to address the issue of the provision of water to all because it has developed a value system that treats water as a commodity. This is particularly the case for the United States, France, and UK. These countries have tossed the provision of water to the dogs of greed that recognize that controlling water resources is in the long term a license for printing money, because consumers are not able to reduce their level of consumption in response to price increases.

European Union (EU) trade negotiators are using the World Trade Organization to gain access to the water services of other countries for the benefit of their own private water industries. The approach to developing countries is to "open your water sector irreversibly for decades. We will buy your bananas and shirts; alternatively you will not get a loan or pay off your debt unless you allow the market to have access to your water services." To date the track record of these companies is to raise tariffs beyond the reach of the poor. In the city of Cochabamba, in 1999 the World Bank imposed privatization of the public water system as a condition of aid for water development in Bolivia. A 40-year concession was granted to the U.S. company Bechtel in 1999. Rates increased by 35 percent, there were strikes and rioting, and people were killed. In 2001, the contract was voided by Bolivia, Bechtel departed, and water was returned to public ownership. Bechtel claimed $50 million compensation. In a case filed before the World Bank trade court, Bechtel abandoned the case after four years of worldwide public protest.[22] There have been violent protests over similar issues in many other developing countries.

There is some recognition that water should not be a commodity controlled for profit. The U.S. Agency for International Development has created a public-private partnership with $4 million from the U.S. government and $37 million from the private sector to provide water in West Africa. Jeffrey Sachs, economist and Kofi Annan's special advisor on the UN Millennium Development Goals, comments:

I am all for the private sector, but the private sector is not going to provide water in West Africa. The private sector wants to make money but you cannot make money off dying poor people. These public-private partnerships are a myth until the public sector puts in the money.[23]

The public sector is presumably the liberal democracies. It seems likely that the consequences of privatization in developing countries will defeat the Millennium Development Goal of halving the proportion of people without safe drinking water and sanitation by 2015. Indeed in a review of progress in 2005, Sachs indicated the goals were "wildly off track." To meet the goals, developed countries had committed to increase aid to 0.7 percent of national income. Currently the richest country, the United States, gives 0.16 percent.[24]

There lies the crux of the issue: The ideology pursued by liberal democracies places matters of world equity, human rights, and ultimately security in the hands of profit-making enterprises. The World Water Development Report from 23 UN agencies has reported that "water resources will steadily decline because of population growth, pollution and expected climate change. This crisis is one of water governance, essentially caused by the ways in which we mismanage water. Attitude and behavior problems lie at the heart of the crisis" with "inertia at leadership level."[25]

THE LAST (FISH) SUPPER

Humanity sees the sea as a source of never-ending riches. Unlike the land, which deteriorates and blows away with misuse, the results of the exploitation of the sea are hidden until the catch fails. Then if the ecological barrier has been breached, recovery is uncertain.

It is a problem that sea and fishing have deep symbolic significance to the human mind. Christianity, and Judaism before it, appreciates the religious significance of the sea. Thus Moses parted the Dead Sea to allow the ancient Israelites to escape the pursuing Egyptians. The sea swallowed these pursuers, allowing God's chosen people to escape to freedom. Likewise, the sea and fish feature in Christianity and the story of Jesus. Apart from the miracle of feeding the 5,000, the imagery of the fish came to be the symbol of the searching human soul, waiting to be caught by the divine angler. As a dual symbol, the Christian him or herself was also, by their example and testimony, fishers of men, a mechanism by which "the Truth" or "Good News" was told.

The sea has been viewed by humanity as a great unknown, a void and emptiness that exists only to be exploited. Unlike outer space, the ocean is readily reached, and that which is needed to be disposed of can seemingly disappear without a trace. Or, so we thought, until comparatively recent times in human history. This mentality of unlimitedness exists because we cannot see its bottom and also applies to the fruits of the sea—fisheries. That which was thought to be boundless and unknown must surely have a potentially infinite number of inhabitants. "There are plenty more fish in the sea" has been a proverb said to heal every heart broken in romance and every disappointed angler. But our belief that there will be plenty more fish in the sea is a romantic illusion that today confronts the starkness of reality.

Global production from aquaculture continues to grow in terms of quantity and relative contribution to the world's supply of fish for human consumption. However, for the purpose of this discussion we will restrict our analysis to fish caught from the commons of the sea. Aquaculture is beyond the aegis of this commons, apart from the use of wild seed (young sea fish) and fish meal for feed, both of which add to the depletion of the sea.

Fishing has become a trade to feed the rich to the detriment of the poor who need coastal fishing for a vital source of their protein. About 1 billion people have fish as their main source of animal protein, and some small island states depend upon fish almost exclusively. Fish provides 2.6 billion people with at least 20 percent of their average protein intake.[26]

The northern European democracies have led the way in this plunder of the commons. In recent times the seas have been fished relentlessly particularly by the North Atlantic nations, so much so that productive fishing grounds are now depleted. The Atlantic cod once plentiful is now an endangered species, and the Pacific and Atlantic herring fisheries have collapsed. Yet even now, attempts at conservation are inadequate. The fleets of trawlers in the harbors of Scotland, Northern England, Newfoundland, Holland, and France are underemployed on the fishing banks of the North Sea and Newfoundland. They are being replaced by vast deep-sea trawlers that roam the deep waters of the oceans as far as Antarctica to bring fish to the well-fed inhabitants of London, New York, and Amsterdam. These factory vessels use sophisticated surveillance methods to detect shoals of fish and then catch, process, and store thousands of tons of fish. Let us look at some facts and figures.

In 1950, 21 million tons of fish were harvested in the world. By 2005, 133 million tons were harvested, of which 93 million tons caught and 40 million tons farmed. The UN Food and Agricultural Organization (FAO) estimates that 52 percent of stocks are "fully exploited" and another 24 percent overexploited. Seven of the top marine species accounting for 30 percent of all production are fully exploited or overexploited. Stocks are greatly depleted and in need of recovery in the Mediterranean and Black seas and Northeast Atlantic, Southeast Pacific, and Antarctic oceans. As a result the total catch of fish from the sea probably peaked early this century and is now declining.[27] A study from the University of Washington paints an even more somber picture.[28]

The harvesting has now moved on to less desirable fish that live in schools in the open seas. Over the decades the size of captured fish has decreased and marine ecosystems are being changed because of the huge biomass removed from the sea, particularly the large predatory species. Furthermore, there is emerging evidence that the harvesting of large fish from any one species may be counter-evolutionary to that species, by leading to smaller and less fertile populations. This offers an explanation why some populations are unable to recover despite moratoriums.[29]

Unfortunately despite the tightening in regulations, fishing methods have become more destructive due to overfishing and damage to marine ecosystems. Fine mesh nets produced a discard rate of unwanted species of 9 out of 27 million tons of fish in the Northwest Pacific and for every kilogram (2.2 pounds) of shrimps caught, more that 5 tons of other species are discarded.[30] Overfishing is encouraged by overcapacity of vessels with government subsidies for investment. As surface fish become depleted, deep-sea trawling, which can be thought of as equivalent to the clear-cutting of mature forests, is increasing and is destroying seabed ecosystems. There is an increasing toll of seabirds, mammals, and turtles caught and killed by lines and nets. Currently, 19 out of 21 species of albatross are under threat of extinction. In chapter 4 we will see that complex ecosystems suffering continuous assault may collapse to the detriment of all species; this toll of so many species must be controlled. Each year there are scientific reports indicating potential global collapse of the species being fished. In a study published in *Science* in 2006 the rate of collapse of fish resources was increasing despite all the warnings of recent decades.[31]

However the impending crisis is not simply one of overfishing. There are numerous environmental factors acting in concert to threaten the remaining biomass of fish. These relate to humanity's actions on land and in the atmosphere. The breeding and living grounds of coastal fisheries are being damaged. Pollution from coastal cities and agriculture is central is this process. Chemicals, nutrients, sediments, pesticides, and even pharmaceuticals are causing deterioration of marine life. Sea grasses and corals where fish breed and young fish mature are damaged. Nutrients produce algal blooms that are toxic to life and reduce the influx of light. Rivers fail to reach the sea because of dams and irrigation, and their normal load of nutrients used by coastal fish is lost. Mangroves where fish breed are cleared for tourism development and for shrimp farms in Asia. Oil spills damage breeding grounds for decades and are increasing in number. In the open seas, accumulation of toxic substances and heavy metals has occurred from oil drilling and the flow of polluted rivers. In the North Sea, fish are deformed by these pollutants, and in the Gulf of Mexico large areas of the sea are dead from pollution.

Analysis shows that these events have complex causes. As one example, the arrival and spread in the United States of the West Nile virus will be described. This virus is normally confined to parts of Asia and Africa. Mosquitoes transmit the virus from birds to humans, resulting in encephalitis, an often fatal inflammation of the brain. In the summer of 1999 birds at the Bronx Zoo in New York died inexplicably. Then humans began to die from encephalitis. The diagnosis of the problem was very slow because Ronald Reagan, believing in small government, had severely reduced funding of the public health service and it had never recovered. However when it was realized that West Nile virus had succeeded in evading the U.S. quarantine service, there was official panic in New York City Hall. It was decided to eradicate the mosquitoes of New York, and the U.S. Air Force was used to spray pyrethroid insecticide over much of the city. No one felt it was important to note that Hurricane Floyd was roaring up the U.S. coastline, and it was soon bucketing rain onto New York and washing the insecticide into drains. Within a week the lobster industry of Long Island Sound, worth hundreds of millions of dollars, was completely destroyed and hundreds of fishermen were out of work. It was overlooked that the mosquito and the lobster are related arthropods and both highly susceptible to the insecticide. West Nile virus spread across the United States and has caused over 600 deaths, high morbidity, and large health costs.[32] Many such events occur regularly around the world, eating away at the viability of the resources of the commons.

But there is a new and even more potentially destructive power at work. Climate change, with increase in sea level and water temperature, threatens marine ecological communities by means of a range of mechanisms already recognized scientifically. Fish have to move habitats because of rising temperatures; corals, the breeding grounds for fish, are bleaching and dying because of the rise in temperature and sea level; and algae and small organisms to feed young fish are failing to thrive in the Antarctic Ocean. The Antarctic Ocean is becoming more acidic due to absorption of carbon dioxide, and this affects growth of tiny crustaceans, the food source for fish.

Finally, there are humanity's overtly destructive methods of making a buck from the increasing shortage of fish, such as the use of cyanide and dynamite on reefs that destroys the entire ecosystem. Another destructive method is the proliferation of fish farms with their increasing pollution of coastal waters and the use of small fish caught from elsewhere for feeding. In Asia, extensive clearance of mangroves to create shrimp farms has both reduced fish breeding areas and exposed the coast and communities to destruction by tsunamis like that of 2004. A distinction must be drawn between this marine fish and shrimp farming, and the rapidly increasing freshwater aquaculture in Asia and particularly China, where it is an important nutritional development.

EQUITY, CONSERVATION, AND THE LIBERAL DEMOCRACIES

The liberal democracies have paid lip service to the fishing crisis but have never done enough to avert it. Indeed, as Anne Platt McGinn and colleagues have pointed out, "the developed countries are contributing to a tragedy of the commons."[33] Every fisherman has been left to make the decision to catch more fish, and liberty has allowed more fish to be caught. As a consequence, northern waters are severely overfished. Then, through water and fishing-access agreements with southern countries, northern countries are essentially transferring their overcapacity of fishing boats. Through these access agreements, northern fishing fleets now take about half of Africa's marine fisheries. The rent typically received by these African countries is less than 10 percent of the value of the catch. The EU paid $426 million to the Mauritanian government to fish in its territorial waters between 2001 and 2006, but already the stocks have been badly depleted.[34] Senegal was forced to open its waters by the International Monetary Fund and World Bank structural settlement programs. It sold licenses to the EU for $19 million per year. Fish are caught and delivered elsewhere by EU fleets without paying taxes, resulting in a dramatic decline in some species of fish.[35] The need to generate foreign exchange to service debt has locked these nations into such exploitative agreements. Thus to feed an increasingly affluent north, fish are taken from the seawaters at minimal cost, with little concern for local environmental regulations, the reduction in employment, or per capita supply of food for poorer nations. Neither is the debt problem resolved because of its size and compound interest.

The demand for protein-based food, promoted by the affluence of the West, fuels this growing desire for increasing quantities of fish. This is not merely because of the direct consumption of fish by humans. About one third of the world's fish harvest is used in animal feeds to produce animal protein. This fetish for animal protein in the developed nations has little to do with sound health and nutrition and much to do with marketing strategies in consumer culture. And as we have already seen, mass consumer culture is a direct product of the individualism, hedonism, and greed of liberal democracy. Fisheries are a clear example of the tragedy of the commons in operation: how unconstrained individual greed leads to the ruin of the resource and condemns previously self-sufficient coastal communities in developing countries to malnutrition.

The main marine fishing nations in order of production are China, Japan, the United States, the Russian federation, Peru, Indonesia, Chile, and India. Clearly the problem is now one of raiding the commons by all who can and not just the responsibility of the wealthiest nations. The liberal democracies display a failure of example that belies their confidence that democracy will be a savior of humanity. The nations of the world have developed regulations through UN auspices but their enforcement always lags the need to do so. The problem of overfishing was recognized by the first FAO Fisheries Technical Committee in 1946 and was flagged in each successive FAO fisheries conference. It is the eternal problem of anthropocentric management of the environment. In the comment of the FAO on management of fisheries, "the rate of real change in management has been slow, and it is debatable whether improvements have kept pace with the increasing pressures on resources."[36] In the diplomatic language of the UN this means that we are losing the resource. As stated in the FAO 2004 report:

The depletion of stocks contravenes the basic conservation requirement of the 1982 UN Convention on the Law of the Sea and of sustainable development. It is also contrary to the principles and management provisions adopted in the 1995 FAO Code of Conduct for Responsible Fisheries. It affects the structure, functioning and resilience of the ecosystem, threatens food security and economic development, and reduces long-term social welfare. The demand for fish as human food may reach around 180 million tones by 2030 and then neither aquaculture nor any terrestrial food production system could replace the protein production of the wild marine ecosystems.[37]

There are intergovernmental organizations with responsibility for managing the high seas, but regulations are only as strong as the political will of the participating organizations; and in Western culture, conservation ranks below profit. Democratic nations have fostered treaties, laws, and regulations but manipulate or ignore them when political or social pressure is applied. The plunder has been extended by the poor and developing nations, often as a necessity, and there is no end in sight but an empty net.

The previous chapter considered the ecological ramifications of climate change and oil use and depletion, and the next chapter will consider the related questions of biodiversity destruction and population growth. The standard response to this material will no doubt be that these problems are ultimately in human nature rather than liberal democracy, perhaps in greed and desire for growth and profit. Now it is not our argument that liberal democracy in the sole *cause* of the environmental crisis. Rather it is that the institution of liberal democracy fails to adequately address the challenges of the environmental crisis, and by giving an even greater license to greed and individual self-satisfaction, it is potentially a more environmentally destructive social system than most other systems under which humans have lived. After all, it is this system that is the only game in town, as its champions tell us. Yet the destruction of the environmental commons is accelerating. The next chapter will further illustrate this thesis by a consideration of biodiversity and population issues.

NOTES

1. The Universal Declaration of Human Rights, Article 25, adopted and proclaimed on December 10, 1948, by the General Assembly of the United Nations.

2. The UN World Water Development Report, *Water for People, Water for Life 2003,* at <http://www.unesco.org/water/wwap/wwdr/table_contents.shtml>.

3. 4th World Water Forum, *Water Supply and Sanitation for All,* p. 109, at <http://www.worldwaterforum4.org.mx/uploads/TBL_DOCS_80_11.pdf>.

4. Ron Neilsen, *The Little Green Handbook. A Guide to Critical Global Trends* (Scribe Publications, Melbourne, 2005), p. 217.

5. A.J. McMichael, *Human Frontiers, Environments and Disease* (Cambridge University Press, Cambridge, 2001).

6. P. Chatterjee, "Dam Busting," *New Scientist,* May 17, 1997.

7. J. Vidal, "Mighty Mekong Close to Rock Bottom," *Guardian Weekly,* April 15–21, 2004.

8. Neilsen, *The Little Green Handbook,* from note 4.

9. L.R. Brown, et al., (eds.), *State of the World 1998* (Earthscan Publications, London, 1998), pp. 5–6.

10. Lance Endersbee, "Australia's Artesian Basin—$14 Billion Down the Drain Each Year," August 15, 1999, at <www.onlineopinion.com.au>.

11. "Dams Control Most of the World's Large Rivers," *Environment News Service,* April 15, 2005.

12. The Report of the World Commission on Dams, *Dams and Development: A New Framework for Decision-Making* (Earthscan Publications, London, 2005).

13. "Dams Control Most of the World's Large Rivers," *Environment News Service,* from note 11.

14. The Report of the World Commission on Dams, *Dams and Development: A New Framework for Decision-Making,* from note 9.

15. Arundhati Roy, *The Algebra of Infinite Justice* (Flamingo, London, 2004), p. 52.

16. Ibid., p. 85.

17. Ibid., p. 70.

18. Ibid.

19. "Who's Behind China's Three Gorges Dam," *Probe International,* at <http://www.probeinternational.org/pi/documents/three_gorges/who.html>.

20. World Watch Institute, Live Online Discussions, "Managing Water Conflict and Cooperation," June 9, 2005, at <http://www.worldwatch.org/live/discussion/109>; J. Reid, "Water Wars: Climate Change May Spark Conflict," *The Independent Online,* Edition 2, March 2006.

21. "The Chinese Miracle Will End Soon," interview with China's Deputy Minister of the Environment, *Dr. Spiegel,* March 7, 2005.

22. J. Luoma, "The Water Thieves," *The Ecologist,* March 2004, pp. 52–57; "Bechtel Surrenders in Bolivia Water Revolt Case Engineering Giant Sought $50 Million, Settles for Thirty Cents," Common Dreams newswire, January 22, 2006, at <http://www.commondreams.org/news2006/0119–12.htm>.

23. "Wealthy Nations Fail to Find Clean Water, Health," *Environmental News Service,* May 30, 2003.

24. J. Sachs, "No Time to Waste," *Guardian Weekly,* September 16–22, 2005.

25. Food and Agriculture Organisation (FAO), *The State of the World Fisheries and Aquaculture 2004,* at <http://www.fao.org/documents/show_cdr.asp?url_file = /DOCREP/007/y5600e/y5600e00.htm>.

26. Ibid.

27. "Depleted Fish Stocks Require Recovery Efforts," at <http://www.fao.org/newsroom/en/news/2005/100095/index.html>.

28. D.L. Alverson and K. Dunlop, *Status of World Marine Fish Stocks* (University of Washington School of Fisheries, University of Washington, 1998).

29. Natasha Loder, "Point of No Return," *Conservation in Practice,* vol. 6, no. 3, July–September 2005.

30. Anne Platt McGinn, Christopher Flavia, and Hilary French, "Promoting Sustainable Fisheries," in L.R. Brown (ed.), *State of the World* (Norton, New York, 1998), pp. 59–78.

31. B. Worm, et al., "Impacts of Biodiversity Loss on Ocean Ecosystem Services," *Science,* vol. 314, 2006, pp. 787–790.

32. The Center for Health and Global Environment, *Climate Change Futures. Health, Ecological and Economic Dimensions* (Harvard Medical School, Harvard, 2005), at <http://www.climatechangefutures.org/pdf/CCF_Report_Final_10.27.pdf>.

33. McGinn, Flavia, and French, "Promoting Sustainable Fisheries," from note 30.

34. P. Brown, "Europe's Catch-all Clauses. EU Fishing Fleets are Devastating Stocks in the Third World," *Guardian Weekly,* March 28–April 3, 2002.

35. F. Lawrence, "The Need for Exports has Led to an Intense Focus on Certain Types of Fish. There Has Been a Rapid Decline in Numbers," *Guardian Weekly,* December 9–15, 2005.

36. Food and Agriculture Organisation (FAO), *The State of the World Fisheries and Aquaculture 2004,* from note 25.

37. Food and Agriculture Organisation (FAO), *The State of the World Fisheries and Aquaculture 2004,* from note 25.

— 4 —

Biodiversity, Ecology, and Population

The dread and darkness of the mind cannot be dispelled by the sun-
beams, the shining shafts of day, but only by an understanding of the
outward form and inner workings of nature.

—Lucretius, ca. 60 B.C.

UNRAVELING THE WEB OF LIFE

A paper in the prestigious journal *Science* in July 2005 by a group of environmental scientists representing a wide range of scientific disciplines indicates that land use practices are destroying ecosystems that are vital for global
sustainability.[1] The lead author, Jonathan Foley, commented that "short of a
collision with an asteroid, land use by humans is the most significant impact
on the world's biosphere."[2] Such dire warnings were used to describe global
warming in chapter 2. It is irrelevant to debate which of these two threats
is the greater, for they are synergistic and related to the many consequences
of economic and population growth. In this chapter we find that humanity
possesses the scientific knowledge that the depletion of ecological services
is a threat to survival yet their protection is not a priority for government
action.

Biodiversity is the variety of all life-forms: the different forms of animals,
plants, and microorganisms, the genes they contain, and the ecosystems of

which they form a part. An ecosystem is a community of different species and their interactions in the habitat within which they live. What are ecological services? Ecological science is the study of the ways in which all living things interact with each other and with their environment. All living things exist in this web of life with mutual interdependence for food and other resources. Humans are part of this web of life. Thus an ecological service is the provision of a resource to humanity by other species. Examples are the provision to provide food, fiber, and purified water, degradation of wastes and pollutants, recycling of nutrients, stabilization of climate, protection against flood and storm, and provision of materials for shelter, medicines, and cultural activity. Clearly, therefore, ecosystem services are an integral part of the health and well-being of humanity and need to be maintained in perpetuity.[3]

The remorseless damage to ecological services by the growth economy and the population explosion since industrialization is the final common pathway of the environmental crisis. The biodiversity that provides these services is lost through deforestation and overplanting of crops, leading to loss of soil, erosion, and desertification; overuse and pollution of rivers; urbanization, overfishing; and climate change. Pollution from mining and oil wells, pipelines, and transport is also significant. The habitat of species becomes fragmented by development and replaced by invasive species brought by trade into environments where there are no natural controls; as a consequence food production on land and in coastal waters is compromised. The overall effect of all these events is to reduce the genetic pool of a species and to isolate it into pockets that cannot interbreed. Consequently there is a rapid increase in extinctions.

The importance of biodiversity is recognized in the Millennium Development Goals,[4] which aim to fulfill the UN declaration of 2000 that stated: "We will spare no effort to free our fellow men, women and children from the abject and dehumanising conditions of extreme poverty, to which more than a billion of them are currently subjected." Goal 7 is to ensure environmental sustainability, and within this goal the role of forests is emphasized:

Forests contribute to the livelihoods of many of the 1.2 billion people living in extreme poverty. They nourish the natural systems supporting the agriculture and food supplies upon which many more people depend. They account for as much as 90 percent of terrestrial biodiversity. But in most countries they are shrinking.[5]

In this chapter we will place some emphasis on the ecological role of forests because this role is easily recognized by the reader, but the arguments we put forward apply equally to many other systems: rivers, soils, the oceans, wetlands, coral reefs, and many more. Ecological systems have an inherent strength and ability for repair. Forests can grow again after logging, soils can regenerate after some degree of overcropping, and rivers can recover if their depleted flows are restored. But only to a point. We will develop the argument that all these resources continue to be used to the point of stress and potential collapse by the society in which we live, and this will threaten our survival.

Let us look at a simple example. An ancient forest has valuable wood for harvesting, which can be used sustainably to provide a living for craftspeople. However, it is more profitable in the short term for the owners, private or government, to chop it down, to make wood chips for paper, and then to optimize future production by converting the land to plantation timber. If the forest is retained it continues to provide ecological services. It filters rainfall to provide pure water supplies at no cost to towns and cities, it evens out the flow from rainfall to avoid flood and drought, and it provides a stable source of sequestrated carbon that would be released as greenhouse gas if the forest is felled. It will maintain numerous species of trees and plants that will help provide a sustainable existence for humanity. Forests create increased rainfall and therefore stabilize climates.[6] However, in economic terms, our value system describes a conserved forest as "locked up," implying that it cannot be used for immediate exploitation and the creation of jobs. It is a measure of the values of wealthy Western civilization that more often than not the forest will be felled.

What does the collapse of an ecological system mean? Essentially the system no longer functions and is not available to provide essential roles within the web of life, some of which may be essential to humanity. For example, overcropping and failure to provide natural manures leads to a reduction of the microorganisms that constitute soil and maintain its structure. It is then susceptible to erosion by wind and flood and is lost to further cultivation. A river may die because its flow is reduced by irrigation, and saline water is returned to the river from the irrigated regions. The animal and plant life of the river then dies, thus destroying the ecological mechanisms that purify the water. The culmination of thousands of such events around the world, all of which are reducing biodiversity, is a global ecological crisis. We will argue that the basic philosophy of Western society embodied in liberal democracy is causing this ecological crisis.

But first we must substantiate the existence of the crisis. It is not sufficient to state that soil is blowing away and rivers are dying. Like the issue of climate change we have to extrapolate into the future from existing evidence. We can measure the numbers of easily visible species and show a steady decline in recent decades. The skin of frogs easily absorbs environmental pollutants, and we can regard the frog as the canary down the coal mine. Its demise is a measure of the health of the environment.[7] Of the 5,743 known species of amphibians almost one third face extinction. In 1998 the *New Scientist*[8] reported that about 12 percent of bird species faced extinction and that there has been a massive reduction in the numbers of more common species in countries with intensive agriculture. This reduction was caused mainly by a loss of habitat and the use of chemicals that kill insects. In 2005, large reductions in the numbers of British woodland birds were reported due to climate change and to loss of habitat and insects.[9] Mostly due to encroachment on their habitat by human activities, 23 percent of the world's mammals also face extinction. Amongst these are our closest relatives, the great apes.[10] It is estimated that in 2003 there were 414,000 apes in the wild. Every two days 414,000 humans are born with the requirement of land and fresh water for their survival. All apes are endangered and expected to become extinct within a few generations because their territory is being taken by humans.

Science can use the health and numbers of certain species as a measure of the health of the environment or more precisely of the ecosystem in which they live. Thus the health of fresh water streams is reflected in the numbers of frogs and of woodlands by the numbers and variety of woodland birds. When the health of one particular species is monitored it is referred to as a "sentinel species." For example the slow decline of the sea otter is a key indicator of the degradation of the Californian coast, which is increasingly polluted and infested with pathogens.[11] The ill health or extinction of a sentinel species often indicates the presence of an environment harmful to humans.

Examination of fossil records indicates that the background rate of extinctions amounts to a few species per year. Currently it is estimated that at least one thousand species are lost each year. This loss is being increased by global warming, and it is estimated that by 2050 15–37 percent of all animals and plants will be threatened with extinction by greenhouse emissions continuing at their present rate.[12]

In the past half billion years of vertebrate existence of life on the planet, sudden climate change, meteors, and perhaps other catastrophic events caused five great natural extinctions, in which perhaps two thirds of species disappeared. Today, scientific opinion is that we are in a sixth extinction period, and this is due to human activity. In simple terms the basic cause is illustrated by the calculations of Vaclav Smil.[13] Six billion humans weigh 100 million tons. If we weighed all wild mammals in the world they would probably not reach 10 million tons, and the mass of all domesticated animals would out-weigh all vertebrates twentyfold. Humans and their livestock consume 40 percent of the planet's primary production of edible plants, and the other seven million species manage on the rest. In biological terms, humans have been able to exist in plague proportions by occupying the ecological space of other species and by using the earth's stores of fossil fuel.

THE ECOLOGY OF HEALTH

In chapter 2 we explained the mechanisms by which human health and well-being are threatened by climate change. Climate change acts in concert with overpopulation, pollution, and many other events detailed earlier in this chapter to damage ecological services essential to the survival of life. In 2005, the World Health Organization (WHO) released its report, *Ecosystems and Human Well-Being: Health Synthesis*.[14] It indicates that the declining condition of most ecosystems is unsustainable and likely to lead to irreversible changes. Those people affected adversely by declining ecosystem services are highly vulnerable and ill equipped to cope with further loss of services. They live with declining agricultural yields and inadequate water supplies and are burdened by infectious diseases that will be perpetuated by climate change. The report envisages loss of soils, famine, and conflict so that food security will not be achieved by 2050 and child malnutrition will be difficult to eliminate.

The concept of ecological services and human health and survival is a concept difficult to sell to politicians and leaders who are imbued by development and the creation of jobs. Most leaders can see that the past 50 years have brought substantial gains in health, well-being, and economic development, but they will not recognize that the gains have changed natural ecosystems more quickly and extensively than any comparable period in human history. The educational task of the scientific community is therefore Herculean. As indicated in the Millennium Ecosystem Assessment Report, the causal links between environmental change and human health are multidimensional. Deforestation, for example, may alter disease patterns as well as local and regional climates, potentially affecting the insects that transmit disease. The disruption of ecosystems may lead to the emergence or resurgence of disease, while local factors such as poverty, in combination with the lack of vaccination and other preventatives, may lead to local establishment and transmission of disease. When these events combine with human activities related to globalization (such as international trade and travel) global pandemics can arise, as illustrated already by the development and spread of HIV/AIDS and potentially by the appearance in human populations of other new infectious disease strains, such as avian influenza.

For example, the forest fires in Indonesia, used annually for land clearing, create air pollution in neighboring countries and respiratory illness in the citizens of Singapore. The ecological consequences of this forest clearing in Indonesia are obvious, but there are also ecological disturbances leading to disease in Malaysia. The fires drive bats carrying the Nipah virus to Malaysia. There, intensively farmed pigs become infected and the virus crosses to humans.

These scenarios illustrate that every event links with others to have biological outcomes that sometimes cannot be predicted. This is not surprising to ecologists who recognize the complex relationships involved and that humans are part of the earth's ecological systems. A.J. McMichael has written that "the health of populations is primarily a product of ecological circumstance: a product of the interaction of human societies and the wider environment, its various ecosystems and other life support services."[15] Once more we are driven to recognize that the imperatives for health cannot be delivered under society's present value system. How have we arrived at this situation?

THE GENESIS OF THE APOCALYPSE

This impending crisis was predicted many decades ago. In The Historical Roots of Our Ecological Crisis,[16] Lynn White observed the historical role of humanity in species extinction, for example, those due to hunting and other subsistence activities. But after 1850 the marriage of science and technology conferred enormous power over nature. White relates this use of power to the democratic revolutions that occurred at the same time and asks whether "the democratic world can survive its own implications. Presumably we cannot unless we rethink our axioms."[17] The answer from the intervening four decades, during which the ecological crisis has become much worse, is clearly that we cannot rethink. White maintains that this inability to act is conditioned by beliefs about our nature and destiny—that is, by religion. In the victory of Christianity over paganism, placation and respect of the spirits of the natural world were replaced by an indifference to the natural world. "Despite Darwin, we are not, in our hearts, part of the natural process. We are superior to nature, contemptuous of it, willing to use it for our slightest whim."[18] Perhaps this is why dismissal of the crisis is greatest in the home of resurgent creationism—the United States. God created man in his image, not as a part of nature. Man named the animals and established dominance over them, and God planned this for man's benefit. It is fair to say however that cogent arguments have been made for Christianity not being dominating of nature,[19] and the late Pope John Paul II was supportive of the natural environment (see chapter 8).

There was an alternative Christian philosophy that failed that of Saint Francis of Assisi, who espoused humility and the equality of all creatures.

And Saint Francis remains a patron saint only for ecologists! White believes that our values are now so imbued with Christian arrogance toward nature that our destiny depends on a religious conversion. We certainly agree that a conversion in thought and values is necessary, but we do not debate further how much of the crisis is attributable to Christianity.

Again the conclusion has to be that a change in philosophy from today's values, imbued through liberal democracy, is essential. In essence, L. J. Perelman makes similar comments about the necessary transition to renewable energy: Society needs to change its values.

Before the short-lived age of industrialization and exponential economic growth, societies maintained their integrity over long periods of time, and will be able to do so again after this age's imminent passing. The glue that has held society together during the Industrial Age is secular, mundane and material. The traditional social glue which will be restored in the course of the coming transition is sectarian, transcendental and spiritual.[20]

Today, theologians like Karen Armstrong[21] continue to ruminate about the need for a holistic vision for society, one that recognizes that all of us equally share a divine biological life and that to damage it endangers the whole. Such beliefs were part of religious ritual and ethical practice in ancient times and are still present in some indigenous peoples who sustain their lands. Try and preach this to the corporate board whose success depends on the exploitation of the commons.

POPULATION AND THE ECOLOGICAL FOOTPRINT

The damage to biodiversity and the ecological services that it provides is caused by the increasing consumption of resources by each individual and by the activities of increasing numbers of individuals on the planet. The UN predicts that the present world population of 6 billion will increase to 8.9 billion by 2050. However this figure may be an underestimate because of the very young age structure of the current world population.[22] The world's population may not stabilize until it has reached 12 billion. By using the "ecological footprint" it is possible to estimate the effects of these increases in population.

The ecological footprint is defined as "the total area of productive land and water required on a continuous basis to produce all the resources consumed, and to assimilate all the wastes produced, by that population wherever on earth that land is located."[23] Presently, there are 9 billion hectares (22 billion acres) of productive land. If this is divided equally there are 1.9 hectares (4.7 acres) for each person on earth. Yet each person requires

2.5 hectares (6.2 acres) for a satisfactory standard of living. If everyone has 2.5 hectares (6.2 acres) the earth will support 3.5 billion people. But it already has 6 billion! How can this be? Many poor populations have a small footprint, for example 0.47 hectares (1.2 acres) for each Mozambican. They are undernourished and are degrading their environment to stay alive. By contrast Americans and Australians have a footprint of 10 hectares (24.7 acres) per person. They are obese, and their bellies and profligate lifestyle are ravaging the environment of other nations. Overall, therefore, the world's population of 6 billion is eating into the earth's ecological capital. This raises the question as to how the earth can support the projected 8.9 or 12 billion people. It cannot unless developed countries move toward an economically sustainable lifestyle.[24]

In fact, at the present time, there is sufficient food to feed the world's population of 6 billion; the problem is one of improper distribution that leaves nearly 1 billion malnourished and susceptible to disease. This 1 billion damages the environment in their effort to survive. The affluent peoples also damage the environment by imposing upon developing countries systems of farming that are socially and environmentally inappropriate. Thus the damage to ecological services results from all the activities of 6 billion humans, rich and poor.

In the case of croplands, the International Food Policy Research Institute estimates that 1.3 percent of croplands per year are damaged from erosion and salination. Erosion of topsoil measured in metric tons (2,205 pounds) per hectare (2.5 acres) per year is 10 in the United States, 40 in China, and probably higher in Africa and India. If we relate available cropland to population, in 1960 there was 0.5 hectares (1.2 acres) of cropland for each person; in 2004 it was 0.23 hectares (0.57 acres). China has 0.08 hectares (0.2 acres) of cropland per person and imports grains. The growing export of crops by modern techniques can be regarded as a form of mining topsoil that is not sustainable, and productivity has been maintained by expanding arable land by clearing forests. As described in chapter 2, present-day productivity is highly dependent on nonrenewable energy provided by oil.

There is another important factor to be considered in the production of food. In chapter 3 we described the rapidly depleting stores of aquifer and ground water and the profligate use of water in developed countries compared to poor countries. Add to this the advent of climate change with more droughts and increased evaporation, and it becomes clear that food production will soon be limited by water resources. Irrigated farming uses two thirds of all water from the depleted rivers and underground reserves. The foods grown are usually high-yielding varieties that utilize large volumes of water. These were developed to yield more food from a given area of land without considering the water requirement. It is estimated that a hamburger uses 11,000 liters (2,860 U.S. gallons) of water, a kilogram of rice 2,000–5,000 liters (520–1,320 U.S. gallons), a jar of coffee 20,000 liters (5,200 U.S. gallons), and the fodder to create 1 liter (0.26 U.S. gallons) of milk requires 4,000 liters (1,040 U.S. gallons).[25] Indeed it is possible to use a footprint for the use of water in the same way as we describe a footprint for land. We will find that soon there will be insufficient water to produce the food needed by the growing population of the world.

CONSUMING THE FUTURE

The linking of biodiversity and population in this chapter to the environmental crisis is because together they represent the final common pathway of damage to the earth by humanity. It is the pressure of an increasing population for more land, water, and food and the pollution, degradations, and climate change that their activity produces that is creating the crisis. There is a complex cumulative impact on other species from every human activity: urban sprawl leading to pollution of rivers by storm water; human settlement in sensitive ecological regions leading to the introduction of domestic animals, pets, and weeds; the use of garden pesticides; dish soap; land contamination from landfills; view management (cutting down of trees); and a thousand other examples.

Two fundamental questions need to be asked. Since population expansion is a key factor in the crisis, what have the Western democracies done about this? Since the crisis is linked to a society based upon consumption, what have governments and corporatism done to recognize the problem and mitigate it?

Today most liberal democracies, apart from the United States, have a birth rate below or at replacement level. The population surge will come in the developing countries. Reduction in birth rates occurs with education and development, and we can ask whether there has been adequate assistance to make this transition to a sustainable form of development, but not to the Western form of development, for this is not possible if greenhouse emissions are to be contained. This transition has not occurred due to the nature and inadequacy of assistance. In addition, most nations have not lived up to commitments made at the UN Cairo conference in 1994 to fund reproductive health. Under the Bush regime U.S. aid to developing countries for family planning has been reduced. By contrast, the most aggressive stance on population growth was made by a totalitarian state, the People's Republic of China, which instituted a one-child policy in 1979. Aggressive policies in democracies fail. For example the vasectomy program in India involved coercion; it was directed at poor people who did not understand the procedure and there was little compensation.[26] The program was a factor in the demise of the Gandhi government in 1977. However, there are competent family planning programs in many countries in Asia and Latin America, but they lack funding for the education and care that is needed to make them effective. The increase in the world's population is a threat to humanity, and the wealthy democracies have not prioritized the need to fund family planning services.

The culpability for the loss of ecological services rests firmly with the liberal democracies and the alliance between government and corporatism. The profit motive now dominant in society is so protected by government that it is difficult to see any remedy under the present economic system. Environmental destruction remains normal practice when profit is threatened. We will give two examples but thousands could be documented. One example is from the developed world and the other from the developing world.

In Australia, a society of medical doctors concerned with the effects of environmental damage on human health wrote to banks and financial institutions about their funds being invested in companies that were responsible for the clear-cutting of old-growth forests in Tasmania and other states in Australia. The scientific and medical reasons for their concerns were clearly outlined. Each reply stated, "Funds management businesses have a legal duty, under the Corporations Act to put their clients' interests first at all times." This is the fundamental problem created by liberal democracy. Profit has precedence over ecological services, climate change, and conservation. Governments and corporatations directly responsible through their acceptance and use of national corporation laws and the international financial institutions bear responsibility by avoiding national restrictions on their activities and by promoting international institutions that demote environmental considerations. These matters will be discussed in detail later in this text.

In the second scenario, the promotion of export-led growth (usually to pay off debt) and reduction of government spending, including that to environmental programs, has been instrumental in increasing deforestation in 15 developing countries.[27] Most studies of deforestation are complex, as can be seen from a study of the Amazon rain forest, which comprises two fifths of the world's remaining tropical rainforest and contains a considerable number of species essential to civilization. Its destruction has a grave danger in causing rainfall reduction in North America and is a major cause of global warming. In the past century, 14 percent of the Amazon forest has been felled, but only 5 percent is expected to remain in a pristine state; 42 percent will be lost or heavily degraded in 20 years.[28] An area of forest the size of Israel is cleared for farmland every year. Pressure to clear this forest comes from agribusiness for the production of soy and beef needed to increase exports to pay debt. The clearing of forest for pasture is fuelled by increasing demands for beef in Russia, China, and countries such as Poland, as well as the United States and Europe. In Brazil the cattle population doubled in the period from 1990 to 2002, and beef exports increased five times between 1998 and 2002. Forest cleared to farm soy, a vital source of protein for China purchased to meet its previous food self-sufficiency, is now eroded by industrial development and population explosion.[29] At the start of this century, Brazil had backed $40 billion of development in 10,000 kilometers (6,200 miles) of highways and in dams, mines, gas and oil fields, canals, ports, and logging concessions.[30] More recently President Luis Inacio da Silva's government committed to spending $135 million on activities to control deforestation, provide better planning, and prevent illegal occupation of land.[31] But the need to preserve forest is further threatened by a major program to produce sugar cane for fuel alcohol.

The complexity of this situation is much greater than described above. Brazil is a poor country seeking to develop. It is under financial pressure from the institutions of the liberal democracies and the free trade ethos. It sees agricultural production as a means of increasing its immediate standard of living. Is it reasonable for wealthy developed countries to prescribe retention of the Amazon when their own development was based on clearing their own forests to fuel agriculture that underpinned their development?[32] In an ideal world with an ethos to protect its climate and biodiversity, rich countries would pay Brazil to protect the Amazon and the proceeds would bring about land reform and value-adding on already cleared land. The world would have less beef, an environmental advantage in itself; the rich would pay more for it and would certainly be healthier. Brazil is engaging in a process of development supported by the culture and economics of liberal democracy for this is the only way it can conceive of improving its lot. It has no alternative under a globalized financial system.

The question arises, if Brazil was able to retain its forests would it be economically advantaged as well as environmentally advantaged? This can be answered in part from the review by Ron Nielsen.[33] When the ecological services provided to humanity are estimated they average $31 trillion per year, which is roughly equal to the global income generated from using natural resources. This establishes that the contribution from nature is enormous. We can then ask, if we convert a natural environment that is maintained with minimal interference into human-engineered systems such as agriculture or fisheries, do we lose or gain? The conversion brings serious economic losses. There is little doubt that the loss of the Amazon forests is an environmental disaster for all humanity and will present an overall economic loss for Brazil.

Forest Trends, a nongovernment organization based in Washington, DC, has analyzed the destruction of rainforests in Papua New Guinea, which contain one of the highest concentrations of biodiversity in the world. Companies based in Malaysia evade the law to cut round logs, which are processed in South Korea, Japan, and China en route to Western countries. The government of Papua New Guinea is implicated. Logging companies are "allowed to ignore PNG [Papua New Guinea] laws and in fact gain preferential treatment in many cases, while the rural poor are left to suffer the social and environmental consequences of an industry that operates largely outside the regulatory system."[34] Apart from China, the countries involved in this destruction are committed to democracy, and they host the companies involved. Both Japan and China recognize the adverse consequences of forest destruction in their own countries yet participate in logging in poorer countries. Over recent decades many recommendations have been made to control illegal logging. All have failed and failure is likely to continue without a fundamental change in value systems and leadership.

The role of democracy and development in the loss of forests is well illustrated by the studies of Neal Englehart[35] on deforestation in eight countries in Southeast Asia between 1990 and 1995. In the two most democratic countries, the Philippines and Thailand, deforestation was the greatest, next were Malaysia and Cambodia, and the lowest rates were in the most authoritarian countries, Burma, Vietnam, Laos, and Indonesia. Now, the leaders of these four regimes are not noted for their environmental credentials, and we cannot conclude that authoritarianism protects forests. However we can conclude that democracy allows greater economic freedom that leads to deforestation. This freedom also allows the community to endeavor to protect the forests, but in every democratic country this opposition is puny in the face of powerful corporate interests. As Englehart observed in Thailand, "privately mobilized capital distributed among a large number of competitors has created highly efficient firms that very efficiently destroy forests."[36] We are not using this argument to recommend the form of authoritarianism present in Southeast Asia; we are indicating that liberal democracies as governance systems are unable to protect the resources of the commons. We will expand this discussion in the next chapter.

The performance of many liberal democracies in sustainability, despite their enactment of environmental protection laws, has been little better than that currently seen in countries with civil disorder and authoritarian rule, where forests and wildlife are plundered for short-term profit or to buy arms. Laws can be downgraded as we see with the Bush regime or challenged legally by corporate entities. Liberal democracy is just that, a democratic freedom to plunder, built on culture and ideology.

In conclusion therefore, society has three options. Firstly, continue as usual; this option will include further refinement of market forces by placing a monetary value on ecological services and developing green accounting and trading schemes to establish rights over the commons.[37] Recognizing the increasing failure of market systems driven by the motivation of greed to properly consider the environment, we have to ask if these developments are capable of stemming destruction. Secondly, democracies must recognize that they have promoted an anthropocentric society that must change to an ecocentric one, considering all that this entails for reeducation, current wealth, growth economics, vested interests, and political power. Realistically, are the democracies likely to choose this option? No, but it may happen eventually as a response to tragedy and a new world order. The third option is the strict enforcement of ecological requirements by government. This recognizes that ecological services have little chance of surviving without tight control by law of all human activity affecting the environment. This option would be thought of as totalitarian by today's free societies, but this may be the only solution for us.

JAPAN TO THE RESCUE

In chapter 8 we ask whether authoritarian technocratic rule, by imposing necessary solutions, could arrest the earth's ecological decline. In history there are examples of environmental decline that threatened the very nature of civilization, being reversed by determined authoritarian rule. In his analysis of societies that fail or survive, Jared Diamond[38] describes the reversal of destructive deforestation in Japan by determined authoritarian rulers. In the mid-seventeenth century, Japan became peaceful, prosperous, and self-sufficient after decades of civil war. The population and the economy exploded, greatly accelerating the cutting of timber used to build houses, castles, and ships, as a fuel for homes and industry, and as mulch for crops. The hereditary rulers, the shoguns, recognized the environmental consequences of erosion and the need to arrest the decline of a rapidly diminishing resource. They saw a threat to the very fabric of their civilization and promulgated a series of complex measures of reforestation in Japan over the subsequent 200 years. Elaborate systems of woodland management were introduced and policed by magistrates and armed guards. Forests became a commons system sustainably managed for the benefit of each village community by issuing separate leases for each household. Guard posts on highways inspected transported timber to ensure observation of rules, and all timber was graded and allocated for specific purposes to avoid waste. The science of silviculture was born and was facilitated by uniform institutions and methods over the entire county. All this was achieved by authoritarian rule in a peaceful society. It is tempting to contrast these events with those in some liberal democracies, for example Tasmania, where all the stakeholders in the natural forests, government, industry, and workers, have united to pillage the forests against the long-term interests of the world community.

What lessons can we learn from the reforestation in Japan? As Diamond points out, these visionary actions were carried out in a society that became destructive to environments outside Japan, so it was not that Confucianism influenced them. Perhaps because there was a recognition of self-interest, for timber was recognized as being of vital importance and also because the hereditary rulers recognized the importance of protecting the needs of future rulers, their offspring. This is not to say that leaders recognizing long-term stakes do not succumb to short-term profits, this having become a hallmark of the democratic leader. But it raises the question as to whether Japan's recovery could be accomplished today under liberal democracy. Perhaps the really big decisions that are vital to the future of humanity are best imposed, and we need to look toward a form of governance that can do this. Hence our assertion that climate change will determine the future of liberal democracy. This is not to deny that bottom-up democratic management of environmental resources is unimportant in some circumstances, and Diamond cites numerous examples that have developed over time and are in use today. Interestingly they encompass microcosms of governance in small rural communities in Swiss alpine villages and in Spain and the Philippines.

NOTES

1. J. A. Foley, et al., "Global Consequences of Land Use," *Science,* vol. 309, 2005, pp. 570–574.

2. Ibid., p. 570.

3. Simon Hales, et al., "Health Aspects of the Millennium Ecosystem Assessment," *EcoHealth,* vol. 1, 2004, pp. 124–128.

4. Millennium Ecosystem Assessment (MA), *People and Ecosystems: A Framework for Assessment and Action* (Island Press, Washington, DC, 2003).

5. Ibid.

6. K. McGuffie and A. Henderson-Sellers, "Stable Water Isotope Characterization of Human and Natural Impacts on Land-Atmosphere Exchanges in the Amazon Basin," *Journal of Geophysical Research (Atmospheres),* vol. 109, November 2004.

7. Steve Connor, "The Polluted Planet: Alarm as Global Study Finds One Third of Amphibians Face Extinction," *The Independent,* October 15, 2004.

8. O. Tickel, "Paradise Postponed," *New Scientist,* January 17, 1998.

9. Robert J. Fuller, et al., "Recent Declines in Populations of Woodland Birds in Britain: A Review of Possible Causes," *British Birds,* March 2005.

10. T. Radford, "Touching the Void," *Guardian Weekly,* August 6–12, 2004.

11. Gary M. Tabor and A. Alonso Aguirre, "Ecosystem Health and Sentinel Species: Adding an Ecological Element to the Proverbial 'Canary in the Mineshaft,'" *EcoHealth,* vol. 1, 2004, pp. 226–228.

12. Chris D. Thomas, et al., "Extinction Risk from Climate Change," *Nature,* vol. 427, 2004, pp. 145–148.

13. Vaclav Smil, referred to in Fuller, et al., "Recent Declines in Populations of Woodland Birds in Britain: A Review of Possible Causes," from note 9.

14. A report of the Millennium Ecosystem Assessment, *Ecosystems and Human Well-Being: Health Synthesis* (World Health Organization, Geneva, 2005).

15. A.J. McMichael, *Human Frontiers, Environments and Disease: Past Patterns, Uncertain Futures* (Cambridge University Press, Cambridge, 2001), pp. xiv–xv.

16. L. White, "The Historical Roots of Our Ecological Crisis," *Science,* vol. 155, 1967, pp. 1203–1207.

17. Ibid., p. 1205.

18. Ibid., p. 1205.

19. R. Bertel, K. Dyer ,and B. Gray, "Is Christianity Green? A Critique of Some Accepted Views of the Relationship Between Christianity and Environmentalism: A Discussion Paper" (Mawson Graduate Centre for Environmental Studies, University of Adelaide, Australia, 1995).

20. L. J. Perelman, "Speculations on the Transition to Sustainable Energy," *Ethics,* vol. 90, April 1980, pp. 392–416.

21. K. Armstrong, "Old World Order Redux," *Guardian Weekly,* December 23–January 5, 2006.

22. D. Pimentel and A. Wilson, "World Population, Agriculture, and Malnutrition," *World Watch Magazine,* Sept/Oct 2004, pp. 22–25.

23. M. Wackernagel and W. Rees, *Our Ecological Footprint. Reducing Human Impact on the Earth* (New Society Publishers, Gabriola Island, BC, Canada, 1996).

24. Ernst von Weizsacker, B. Amory Lovins, and L. Hunter Lovins, *Factor 4: Doubling Wealth-Halving Resource Use. The New Report to the Club of Rome* (Allen & Unwin, Sydney, Australia 1997).

25. J. Vidal, "Running on Empty," *Guardian Weekly,* September 29, 2006.

26. Susanne L. Cohen, *Vasectomy and National Family Planning Programs in Asia and Latin America* (University Center for International Studies, Pittsburgh, 1996).

27. American Lands Alliance Report, at <http://americanlands.org/imf_report.htm>.

28. J. P. Silveira, et al., "Development of the Brazilian Amazon," *Science,* vol. 292, June 1, 2001, pp. 1651–1654.

29. Jonathon Watts, "A Hunger Eating up the World," *Guardian Weekly,* January 20–26, 2006.

30. J. Vidal, "Brazil Sets Out on the Road to Oblivion," *Guardian Weekly,* July 19, 2001.

31. D. Kalmowitz, et al., *Hamburger Connection Fuels Amazon Destruction, Cattle Ranching and Deforestation in Brazil's Amazon* (Center for International Forestry Research, Bogor Barat, Indonesia, 2004).

32. B. Holmes, "The Amazon," *New Scientist,* September 21, 1996.

33. Ron Nielsen, *The Little Green Handbook. A Guide to Critical Global Trends* (Scribe Publications, Melbourne, 2005).

34. *Logging, Legality and Livelihoods in Papua New Guinea: Synthesis of Official Assessments of the Large-Scale Logging Industry* (Forest Trends, Washington, D.C., 2006), at <http://www.forest-trends.org/documents/png/>.

35. Neal Englehart, quoted in chapter 4 of James David Fahn, *A Land on Fire: The Environmental Consequences of the Southeast Asian Boom* (Westview Press, Boulder, CO, 2003).

36. Ibid., p. 117.

37. "Rescuing Environmentalism Market Forces Could Prove the Environment's Best Friend—If Only Greens Could Learn to Love Them," *The Economist,* April 23, 2005.

38. Jarad Diamond, *Collapse: How Societies Choose to Fail or Survive* (Penguin Books, New York, 2005).

— 5 —

Democracy Is What You Think It Is

The people of England deceive themselves when they fancy they are free; they are so, in fact, only during the election of members of parliament: for, as soon as a new one is elected, they are again in chains, and are nothing. And thus, by the use they make of their brief moments of liberty, they deserve to lose it.

—*Jean Jacques Rousseau*

PLATO'S SAVAGE BEAST

Democracy is often defended but seldom defined or explained. The most famous definition is that of Abraham Lincoln—government "of the people, by the people, for the people." But "people" can mean some or all of the people, and government can have various degrees of participation by the people. Commonly, government is by elected representatives acting in their own interests but emphasizing that they are governing for all the people. Democracy has become the playground for self-interest and probably always was. Joseph Stiglitz has noted that few believe that one should be able to sell one's vote, but

[i]ndirectly, through the media, votes are bought and sold. People have to become informed, convinced to go to the trouble of voting, even taken to the polling booth, and all of this costs money. That is why campaign contributions are so important. But

individuals and even more so corporations that contribute expect something in return. They are buying government support.[1]

We will argue therefore that democracy is seriously flawed. This was recognized by Plato when democracy evolved in ancient Greece. Plato had an image of the voters as a "savage beast" that had to be pandered to by the beast's keeper, the politician. Elections have become the promise of largesse to the voter. This is a fundamental reason why important decisions on true sustainability are not made today. Once more we remind the reader that unless liberal democracy can successfully address this issue, there is no rationale for its existence.

A perusal of current texts referring to democracy and related subjects illustrates the problems of definition. Each person looks at democracy from their own point of view of society. Democracy is what you think it is, or so it seems. Noreena Hertz, in her book *The Silent Takeover*[2] is concerned about the threat that corporatism poses to democratic institutions, but she never considers why democratic institutions are a good thing and what in any case they are.

George Monbiot, in *The Age of Consent*,[3] hopes to see a new world order arising from a global democratic revolution. He offers this definition of democracy: "a form of government in which sovereign power belongs, in theory, to the people, in which those people have equal rights, and in which the will of the majority is expressed and exercised through elections between competing candidates and parties."[4] Monbiot's definition of democracy is deficient in that it does not tell us how the will of the majority is expressed and exercised through elections. Monbiot, a committed environmentalist, wants a more participatory democracy in the face of evidence that democracy cannot stem global environmental deterioration. In contrast to Plato, Monbiot assumes that human nature can be trusted to retrieve the situation. We dispute that point.

Joseph Stiglitz, Nobel Laureate in economics, equates effective democracy to the function of the market, "if our democracies are to work, citizens must understand the basic issues confronting our societies and the way in which their government works. And no issues are of more importance for most people than those that surround our economy and the relations between the market and the government."[5] Maybe so, but this doesn't tell us what a democracy actually is and why spreading it around the world is worth shedding blood.

FORMS OF DEMOCRACY

Various ways in which democracy could be achieved have been devised. In the system of direct or *participatory democracy* every person votes on every issue, willing every law passed. In *representative democracy* all decision-making power is given to elected individuals. We will consider participatory democracy first. This is the rule of unanimity that is said to create a legitimate society because there is universal consent to its laws. But such a system works only if all members of the voting community unanimously agree on a matter; otherwise by definition the system fails. Only one "nay" defeats the system. It is necessary therefore that after all debate and deliberation there is ultimately unanimous agreement on laws. This system of democracy was developed in ancient Greece and Rome when an assembly of people met, debated, and made decisions. To Plato, this was the tyranny of the mob and sooner or later it would violate human liberty. It is rightly described as "mobocracy." To operate in the complex society of today this form of democracy would need to make use of household computer voting if there was to be any chance of it functioning at all. It is doubtful, to say the least, that in the highly heterogeneous societies in which we live unanimous agreement could be reached on anything.

Nevertheless, elements of primary democracy do exist in some countries. In some of the Swiss cantons there are assemblies of citizens that make decisions, and throughout the Federation of Switzerland it is possible to initiate citizen referendums on government proposals. The Swiss governance system seems effective and stable and will be briefly discussed, for it represents an amalgam of direct and representative democracy. Although the Swiss system is a form of representative democracy, our primary focus in this critical analysis is on the limitations of direct democracy, and we will focus on that aspect of the Swiss system. A critique of representative democracy will be made later in the chapter.

The Swiss system is a federal system, meaning that power is constitutionally divided between the federal and state governments (or in the case of Switzerland, the cantons) so that each authority exercises responsibility for a set of functions within a sphere or jurisdiction of responsibility. In Switzerland the federal government has sovereignty over the cantons on issues not falling under cantonal authority. The Swiss political system, like the United States and Australian systems, has two chambers, a house of representatives and a senate.

What is interesting about the Swiss system is the use of referendums. Cantons have the right, as do parliamentarians, to put forward initiatives to parliament, either as bills or changes to the Swiss constitution. As well, a proposal can be prevented from becoming law if a referendum defeats the proposal. Such a referendum can be initiated if the signatures of 50,000 voters are obtained against the law in the 100 days after the decree is published. Alternatively, if eight cantons oppose the changes, a referendum can be held.

The people's initiative requires a majority of the popular vote as well as the majority of the cantons to be successful. Votes made by the inhabitants of small cantons count more than those of inhabitants of large cantons, because of the principle of *proportional representation*. This is an electoral system that seeks to ensure that minorities are represented in the legislature through distributing seats in accordance with the proportion of the total vote recorded. A nonproportional representation system can disenfranchise nearly one half of the voters in an electorate and, with some voting systems, even a majority. The downside of a proportional representation system is that it could lead to a tyranny of the minority. The small German-speaking cantons in Switzerland, for example, have a veto, so that a proposal could fail even if it was accepted by the majority of the Swiss voters if the German-minority cantons disapprove.

This raises the first paradox of democracy: There is a need to prevent a majority from oppressing a minority, but there is also a need for a system that prevents a minority from oppressing the majority. A minority could hold out against necessary environmental laws for the common good, purely because of selfish ends. Direct democracy appears to be a good idea on paper, but in practice it frequently leads to socially regressive results; for example, Switzerland only gave women the vote in the 1970s.

Let us consider majority rule. According to the *majority rule* principle, when the electorate is divided a vote is taken, one person a vote and majority wins. Most democratic theories involve majority rule, arguing that if the majority had willed the law, then because the law is somehow the voice of the people all must obey it. This system has obvious problems. Why should a *minority* who voted against the law accept it? The usual answer is because the minority accepts the system of majority democracy so they have to put up with the result. In any event the majority could probably rule by force anyway. Furthermore the majority system is the best safeguard against the arbitrary rule by power elites, or so democratic theorists argue.

MAJORITY RULE: THE AYES HAVE IT

In mounting a critique of the majority rule principle, we turn first to the American philosopher Robert Paul Wolff. In his book *In Defense of Anarchism*,[6] Wolff sets out to show that democratic theory fails to offer a sound moral reason for the individual obeying the state. The reason for this is that an individual's autonomy (i.e., personal freedom) is lost. Wolff concludes that one must either reject the value of autonomy or else conclude that all forms of government are not legitimate, which is what the doctrine of anarchism entails and which is Wolff's solution to this problem. He begins his argument by introducing the fundamental problem of politics, the problem of authority. How can the existence of authoritarian entities such as states be reconciled with human autonomy, as states typically restrict freedom of the individual because states assert the right to command and be obeyed? Individuals are free, and yet the state restricts their freedom. Democratic theory attempts to solve this fundamental problem of politics by saying that freedom requires that people will not be subject to others, because if the people rule themselves, then an individual can be both ruler as well as ruled.

It is argued in defense of democracy that minorities should accept the system because democracy advances the welfare of everybody. However, the argument that welfare as a whole is advanced does not effectively distinguish democracy from benevolent dictatorship, which may also advance the welfare of everybody. Replying to Winston Churchill's remark that "democracy is the worst form of government except for all the others,"[7] Wolff says that if this is so, then the citizens of America, Australia, and other Western countries "are as much subjects of an alien power as the Spaniards under Franco or the Russians under Stalin."[8] They are merely more fortunate in their rulers!

Sometimes supporters of majority rule democracy attempt to justify their system by an argument known as the *social contract argument*. The idea here, stated briefly, is that everybody has promised (i.e., contracted) to abide by majority rule. Thus there is an obligation to abide by majority rule. In other words, by some usually unspecified way, we have agreed to abide by majority rule. But have we really promised to obey majority rule? Is it the case that a minority has somehow agreed to accept the decisions of the majority? Surely not as a matter of historical fact. None of us have embraced any sort of explicit "sign on the dotted line" contract with the state or even a verbal contract for that matter. But supporters of democracy say that by living in such a system and reaping its benefits, one is implicitly accepting the majority rule system. You live in this society, you benefit, so you must accept the system. But this is no answer to *why* the democratic system of majority rule as a whole should be accepted. The question still remains: Why should such a system be obeyed if the requirements of a majority decision involve the surrender of individual freedom? It is arguable that the minority is not reaping any benefits, but in fact is having its interests frustrated. The fundamental problem of why we should obey a majority decision is therefore left unsolved.

Further to this point, the social contract argument could be used to justify alternative nondemocratic systems of government and does not therefore uniquely justify democracy.[9] The same style of argument could, for example, be used to justify the obedience of subjects who have been born into a dictatorship.

Wolff also observes that the principle of majority rule faces voting paradoxes that challenge its self-consistency. Much has been written on this topic since Wolff's book was published in 1970, and a brief mention of the field of social choice paradoxes will be made to put a final nail into the coffin of majority rule democracy.

The idea that democracy was inconsistent or self-undermining was first expressed in Plato's *Republic* in his claim that democracy leads to tyranny. The founding fathers of America were also fearful of the possible tyrannical evolution of democracy. Tyranny did evolve from liberal democracy in 1933 in Germany, when 60 percent of the electorate voted into power the National Socialists. In subsequent chapters we discuss the role of corporatism in the creation of Nazi tyranny and the innate tendency of all democracies to become authoritarian because of humanity's innate acceptance of authority.

Democracy did not survive for long in Germany. Karl Popper reflecting on this situation has described a "paradox of democracy":

> put in a hopeless intellectual position [are] all those democrats who adopt, as the ultimate basis of their political creed, the principle of majority rule or a similar form of the principle of sovereignty. On the one hand, the principle they have adopted demands from them that they should oppose any but the majority rule, and therefore the new tyranny [majority rule would install]. On the other hand, the same principle demands from them that they should accept any decision reached by the majority, and thus the rule of the new tyrant. The inconsistency of their theory must, of course, paralyze their actions.[10]

It was the French social mathematician Antoine Condorcet (1743–1794) who first realized that the idea of "the will of the people" or a collective preference of a community by means of a majority voting procedure led to a paradox. To deal with this paradox social mathematicians began to formalize, by mathematics and symbolic logic, the field of social choice, of which majority rule is a part. An axiomatic theory—meaning a theory with a list of clear, listable fundamental assumptions—was constructed to describe what a liberal choice theory was. Thus axiom or assumption number one is *collective rationality,* that the collective preference of a community can be derived from ordering the preferences of individuals. That seems plausible enough because for liberalism, all that exists are individuals. The other three assumptions are more technical and in this text will not be summarized. Nevertheless the assumptions were regarded as intuitively obvious.

The mathematician Kenneth Arrow showed that those very basic assumptions led to a contradiction. No voting method could be adopted by a community to make a choice between a number of options that did not contradict the preferences expressed by individuals. Democracy led to a dictatorship![11] And to rub salt into the wound, this most depressing result was proved by mathematical logic. All attempts to state a plausible liberal theory by reworking the fundamental assumptions have led to other critics devising more ingenious contradictions![12]

All of this is a major problem for liberal democratic theory because it has been shown by rigorous mathematical methods that the theory is inconsistent or contradictory. The principle of majority rule should thus be rejected as a universally valid method of obtaining the democratic theorists' illusive will of the people.

THE PROBLEM OF REPRESENTATION

Democratic theorists believe that the chaos and other defects of direct democracy can be overcome through the majority rule of representative democracy. In this, voting by some or all of the population is used to elect representatives who are proposed by political parties. When elected, the majority party forms the government. The leader is usually the elected leader of a party but in the United States the president is elected directly by the people. This is a system of majority rule. Laws are made by the majority, and all must obey because for various reasons the decisions of a majority are binding on the minority.

When the authors refer to "democracy" in other sections of this book, we are referring to the forms of representative democracy in developed Western nations. We have seen that majoritarian democracy faces severe problems. Now we turn to consider *representative democracy,* the main defect of which is obvious to all. There is a day of democratic voting followed by three or four years of benevolent dictatorship, or perhaps not so benevolent dictatorship, when wars may be declared, environments destroyed, and public assets sold, all in the name of the populace but without reference to the voter. The voter has indulged in a "one-night stand" and then remains democratically celibate for the next three years. Thus, in this system, by means of an election, the people assign all decision-making power to the so-called representative, who is said to represent the will of the people, whatever that means. However representatives typically vote without a specific list of the people's preferences on future voting issues and thus vote on issues as they see fit. This system is said to be different from a limited term dictatorship because the rulers are chosen from the people, by the people, are expected to act in the interest of the people, and are subject to recall and election. In short, we put them there and we can remove them, so the defenders of democracy argue.

In chapter 4, we discussed the importance of retaining the world's forests and the failure of liberal democracies to do so. Let us look at one example;

the most democratic of the liberal democracies that has a meticulous system of proportional representation built into a representative democracy. This is the system in Tasmania, a state in wealthy Westernized Australia. There, it appears that the will of the people is to continue to destroy the mature forests of Tasmania for the export of woodchips. Both major political parties support this endeavor, so the destruction continues regardless of which party is in power. However, opinion polls indicate that a majority of the population wishes to preserve the forests and their viewpoint is supported by a minority Green Party. The Green Party does not gain power because other aspects of its platform do not attract votes and because the major parties use voting preferences at elections to exclude it. In terms of the future needs of the world, in Tasmania representative democracy is the means whereby environmental destruction is planned and executed against the will of the people.

THE TRAGEDY OF THE COMMONS AGAIN

In most cases, however, it is the will of the people, fostered by the individualism of liberal democracy, that treats the environment as a resource. It is now relevant to explain in more detail the thesis of Garrett Hardin in his seminal paper, "The Tragedy of the Commons," published in the journal *Science* in 1968.[13] This paper exposes the defect that makes democracy unsustainable. The commons of Anglo Saxon culture was the pasture open to the cattle of all villagers. Hardin explained:

As a rational being, each herdsman seeks to maximise his gain. Explicitly or implicitly, more or less consciously, he asks "What is the utility *to me* of adding one more animal to my herd?" This utility has one positive and one negative component. The positive component is the increment of one animal. Since the herdsman receives all the proceeds from the sale of the additional animal, the positive utility is nearly +1. The negative component is a function of the additional overgrazing created by one more animal. Since however the effects of overgrazing are shared by all herdsmen, the negative utility for any particular decision making herdsman is only a fraction of minus 1. Each herdsman concludes that it is sensible to add another animal to his herd, and another, without limit... Therein is the tragedy, and in a world that is limited, freedom in the environmental commons brings ruin to all.[14]

The "world commons" is the stability of the resources of land, sea, air, and fresh water necessary for the health and well-being of humanity. We now have a clear vision of the "ruin to all" predicted by Hardin. It is the confluence predicted this century of the above problems, population growth, depletion of resources, and the ravages of climate change. All our problems can be placed in the context of the commons.

Thus we see that it is in the interests of the individual to break the rules that might be made for the survival of all herdsmen and the resource. This individual will behave acquisitively only if he or she knows that everyone else will comply with the rules. The rules must be strong and inviolate to stop conflict between individual rationality and the common good. Even then there will have to be penalties to ensure compliance. Democracy is indicted because it is unable to defend the commons. We find that democratic states behave in the same way as individuals (e.g., European Community [EC] decisions on fishing discussed in chapter 1). Thus both individuals and states act in ways that are individually rational but environmentally destructive. A nation such as the United States may decide to continue polluting the commons with greenhouse gases to the detriment of all other states because it has immediate economic advantage. In the case of the EC it is of short-term advantage (i.e., job stability) to continue fishing despite the recognition that it is unsustainable. Unless this problem can be resolved to preserve a sustainable world there is no case for the continuation of liberal democracy or nation states. There should be one government, and our argument in chapter 8 would make this government authoritarian.

There are additional cogent reasons why the commons cannot be saved, and these will become apparent in the next chapter. They relate to the mutual dependence of liberal democracy and corporatism. Democracy has a façade of environmental laws and protection but when a corporation wants a resource invariably it will get it, laws will be changed, exceptions made, and rules bent for it is in the personal interest of governments of elected representative to keep people in jobs and collect taxes. Decade after decade the encroachments are remorseless.

THOSE AGAINST DEMOCRACY

French political thinker Jean Jacques Rousseau (1712–1778) opposed classical representative democracy with the following argument in the *Social Contract*:

Sovereignty cannot be represented for the same reason that it cannot be alienated: its essence is the general will, and that will must speak for itself or it does not exist: it is either itself or not itself: there is no intermediate possibility. The deputies of the people therefore, are not and cannot be their representatives; they can only be their commissioners, and as such are not qualified to conclude anything definitively. No act of theirs can be a law, unless it has been ratified by the people in person; and without that ratification nothing is law.[15]

In other words there is a fundamental problem with representation itself. Even ideally if the number of issues is large and there are a number of positions that could be taken on each issue, then the permutations of policy platforms will always be greater than the number of candidates. Representation cannot occur.

More recently Gordon Graham in *The Case Against the Democratic State*[16] has developed a devastating critique of the democratic state. He argues that all of the standard arguments justifying states fail. Graham does not accept anarchism, the idea that there should be no state at all, but opts for a form of republicanism with voting retained as a purely expressive activity. He argues convincingly that in a mass society, voting itself is causally impotent and genuine representation an illusion. Graham believes that states are necessary to supply public goods that markets cannot supply.

By contrast, Wolff concludes his critique of democracy by accepting the doctrine of anarchism, that states are unnecessary and unjustified. He argues that the principle of individual autonomy must be accepted because otherwise citizens become no more than children.

Hans-Hermann Hoppe in *Democracy: The God that Failed*,[17] carries the argument of Wolff and Graham through to its logical conclusion. Hoppe is an "anarcho-capitalist," meaning that he believes that the state is unjustified and unnecessary (anarchism) and that capitalist institutions can in any case do the same job as the state—but better. He writes from the perspective of the "Austrian School of Economics," which is a position that sees the free market as solving most economic and social problems.

We have argued against such a world view in this book and will not repeat our arguments again, but we summarize by saying that the anarcho-capitalist's faith in the market mechanism is as naïve as communism's faith in the goodness of human nature.

Anarchists are mistaken in supposing that social order will be maintained without a state of some shape or form. Graham argues that people by nature desire social order and that conflict and crime are minority occurrences. That naïve view of human nature is refuted every time there is a major power blackout in U.S. cities. Looting and crime increase. The social chaos and death in the failed states in the Horn of Africa indicates that in anarchy, "states" will still arise—the rule of warlords.

Hoppe is one of the few anarchists to address this criticism. He supports the idea of an armed citizenry, much like the Wild West, where everybody has a gun on their hip. Insurance companies would encourage gun ownership by offering lower premiums to armed and trained clients. Insurance companies for Hoppe stand in for governments. They will be defensive agencies and will attempt to prevent criminal predatory behavior, presumably by apprehending and punishing those liable for damage. But the problem here is that there are likely to be a plurality of rival insurance companies—as well as mafia-style protection societies—that through rivalry without a state as umpire, will fight it out Wild West style. Hello social chaos. Is this the America of the future?

Although we reject Hoppe's critique of the state, his primary criticism of democracy is sound in our opinion. Democracy leads to social decay because politicians, with some exceptions, are short-term caretakers and career seekers, and they are only focused on the next election. As an economist, Hoppe observes that this leads to massive public debts. We would add the much more fundamental problem—it also leads to the tragedy of the commons problems as well as destruction of the environment.

CONCLUSION

Our position differs from Wolff and other anarchists also insofar as we reject the principle of autonomy, the foundation belief of liberalism. It is the argument of this work that liberalism has essentially overdosed on freedom and liberty. It is true that freedom and liberty are important values, but such values are by no means *fundamental* or *ultimate values*. These values are far down the list of what we believe to be core values based upon an ecological philosophy of humanity: survival and the integrity of ecological systems. Without such values, values such as freedom and autonomy make no sense at all. If one is not living, one cannot be free. Indeed liberal freedom essentially presupposes the idea of a sustainable life for otherwise the only freedom that the liberal social world would have would be to perish in a polluted environment.

The issue of values calls into question the Western view of the world or perhaps more specifically the viewpoint that originates from Anglo Saxon development. It is significant that the "clash of civilizations" thinking espoused by Samuel Huntington, a precursor of the neoconservatives, has generated much debate and support. Huntington's analysis involves potential conflict between "Western universalism, Muslim militancy and Chinese assertion."[18] The divisions are based on cultural inheritance. It is a world in which enemies are essential for peoples seeking identity and where the most severe conflicts lie at the points where the major civilizations of the world clash. Hopefully this viewpoint will be superseded, for humanity no longer has time for the indulgence of irrational hates. The important clash will not be of civilizations but of values. The fault line cuts across all civilizations. It is a clash of values between the conservatives and the consumers. The latter are well described in this book. They rule the world economically, and their thinking excludes true care for the future of the world. The conservatives at present are a powerless polyglot of scientists, environmentalists, farming and subsistence communities, and peoples of various religious faiths, including a minority of right-wing creationists who think that God wishes the world to be cared for. They recognize the environmental perils and place their banishment as the preeminent task of humanity. The fight for minds, not liberal democracy, will determine the future of the world's population. If conservative thought prevails it may unite humanity in common cause and heal the cultural fault lines.

In the next two chapters we will develop further our critique of liberal democracy, arguing first that democracy is already at an end through the undermining of democratic institutions by man's inherent mentality and by global corporate capitalism. We will find that the latter has become Plato's beast and the keeper that panders to the beast has become the democratic government. This is so, regardless of the correctness of the arguments of this chapter. In chapter 7, we will look more closely at liberalism itself and detail its philosophical flaws. This will complete our multipronged philosophical and ecological dissection of liberal democracy. Having exposed what remains beneath the mummy's shrouds, it will remain to search for an alternative system and explore whether liberal democracy can be saved by radical reforms or political surgery or resurrected from the tomb of its self-destruction by divine intervention.

NOTES

1. Joseph Stiglitz, *The Roaring Nineties, Why We're Paying the Price for the Greediest Decade in History* (Penguin Books, London, 2003), pp. 198–199.

2. N. Hertz, *The Silent Takeover: Global Capitalism and the Death of Democracy* (William Heinemann, London, 2001).

3. G. Monbiot, *The Age of Consent: A Manifesto for a New World Order* (Flamingo/Harper Collins Publishers, London, 2003).

4. Ibid., p. 25.

5. Stiglitz, *The Roaring Nineties, Why We're Paying the Price for the Greediest Decade in History,* from note 1, p. xxviii.

6. R. P. Wolff, *In Defense of Anarchism* (Harper and Row, New York, 1970).

7. K. Jasiewicz, "The Churchill Hypothesis," *Journal of Democracy*, vol. 10, no. 3, 1999, p. 169.

8. Wolff, *In Defense of Anarchism,* from note 6, p. 40.

9. Wolff, *In Defense of Anarchism,* from note 6, pp. 42–43.

10. K.R. Popper, *The Open Society and Its Enemies,* vol. 1, 5th edition (Routledge and Kegan Paul, London, 1996), p. 123.

11. See M.D. Resnick, *Choices* (University of Minnesota Press, Minneapolis, 1987); A. Weale, "The Impossibility of Liberal Egalitarianism," *Analysis,* vol. 40, 1980, pp. 13–19.

12. See J.S. Kelly, *Arrow Impossibility Theorems* (Academic Press, New York, 1978).

13. G. Hardin, "The Tragedy of the Commons," *Science,* vol. 112, 1968, pp. 1243–1248.

14. Ibid., p. 1244.

15. J. Rousseau, *The Social Contract*, Bk. III, Ch. 15, 1762.

16. Gordon Graham, *The Case Against the Democratic State* (Imprint Academic, Charlottesville, VA, 2002).

17. Hans-Hermann Hoppe, *Democracy: The God that Failed: The Economics and Politics of Monarchy, Democracy, and Natural Order* (Transaction Publishers, New Brunswick, NJ, 2001).

18. Samuel P. Huntington, *The Clash of Civilizations and the Remaking of World Order* (Simon & Schuster, New York, 1996).

—

`Politics?`

' government, full of variety
uality to equals and unequals

—Plato

ARGUMENT

...y democracy is failing and why
... orthodox liberal democratic politics is at an end. There
are two related arguments. The first is that even if functional democracies
were possible, there is a multitude of forces that are acting to corrupt them
and prevent them from truly representing the voice of the people. Various
powerful elite groups rule the modern liberal democracy, being based in finance, media, business, and the military; and they have their own agenda that
does not have advancing the interests of the common good among them.
Given these structures of power and inequality, democracy is an illusion,
and ordinary people are to be exploited as these elites see fit. The Australian journalist Paul Sheehan in *The Electronic Whorehouse*[1] painted a picture
of Australia that is lied to by politicians and the media in the most blatant
of ways. John Le Carre, the writer and former spy is quoted as saying to
Sheehan: "I think we are dealing with an octopus. Advertising as news. It's

very skillfully done. The methods of seducing the media, the ingenuity of the spin has reached the point where we, as a general public have never been lied to by such sophisticated means as now."[2]

The evidence of decline is present in many ways: the fall in voter interest in political matters except for those issues that directly affect their wallet or pocketbook, the decline in the numbers voting, and the standing of politicians in many Western democracies that resides at the bottom of a trustworthiness scale along with used car salesmen. The instillation of fear and negative advertising has replaced dedicated policies that could offer hope of a sustainable future for society. For those who do think about the issues, there is disillusionment that policy is based upon community polling and by the self-interest of the politicians.

Events in Australia in early 2007 provide a cogent example. Over the 10 years in power the Howard government had expressed significant skepticism about climate change, which in their minds was categorized as an environmental issue to be contained with some suspicion. Government policy was heavily influenced by the fossil fuel industry and alternative energies were relatively neglected. John Howard refused to see Al Gore, who visited Australia to promote his film *An Inconvenient Truth* on climate change, and his industry minister described the film as "entertainment." Then three events changed the situation. Australia was suffering its worst ever drought, and the farming sector, influential in government, began to express conviction that this calamity was related to climate change. A series of opinion polls indicated that a large majority of the electors had serious concern about climate change. Then the Stern Review was released. The polls necessitated catch-up politics, and there was a surge of statements on solar energy and on the need for nuclear energy to provide the technological fix because it was "clean and green." Ministers who were committed skeptics over 10 years suddenly became experts on climate change, its dangers and solutions. A decade of neglect and lack of leadership had been galvanized by community leadership. Of course this government backflip was assisted by the Stern Review, "The Economics of Climate Change,"[3] which described the severe economic consequences of not addressing the issue immediately. It took the issue away from an environmental classification to a more comfortable economic one within the paradigm of liberal democracy.

The second argument to be pursued in this chapter is that based upon Darwinism and the sociobiology of human nature, which indicates that structures of dominance are inescapable. We are stuck with the elites, so let us have the right sort of elites.

CORPORATISM AND GLOBALISM

Throughout this book examples have been given showing how corporate influence on governments determines poor environmental outcomes. In the United States intense lobbying by dedicated people with ready access to government because of financial contributions to election campaigns has thwarted the implementation of new environmental laws, neutered existing laws, and sabotaged international agreements. Environmental laws are seen as surmountable by transferring manufacturing to developing countries, and national environmental regulations are denounced as hindrances to free trade by the World Trade Organization. In all countries corporatism continues to use the environment for externalities. While the public expression of social responsibility is now fashionable and is used by some corporates to emphasize branding, the fundamental philosophy remains unchanged and enshrined in law. As held by Milton Friedman, there is but one social responsibility of the corporation and this is to make as much money as possible for their shareholders.[4] This is a moral imperative, and to choose environmental goals instead of profits is immoral. We believe that this is the rock upon which the leaking ship of democracy steered by Plato's savages will finally founder.

It is important to emphasize that the environment is not the only sector of society to suffer under the corporate yoke. One cynical view of corporatism is that of Arundhati Roy given in the Sydney Peace Prize Lecture, "Peace and the New Corporate Liberation Theology":

the Lazy Managers Guide to Corporate Success, first stock your Board with senior government servants. Next stock the government with members of your Board. Add oil and stir. When no one can tell where the government ends and your company begins, collude with your government to equip and arm a cold blooded dictator in an oil rich country. Look away while he kills his own people. Simmer gently. Use the time to collect a few billion dollars in government contracts.[5]

Indeed, most so-called Western societies are not democracies as such but plutocracies, societies ruled by the wealthy. In this context Franklin D. Roosevelt's comments in the 1930s about the emerging fascist threat is just as relevant today about the corporate actions of an unallocated economic elite who manipulate the life and destiny of humanity. The liberty of democracy is not safe if the people tolerate the growth of private power to a point where it becomes stronger than that of the state itself. That, in essence, is fascism: ownership of government by an individual, by a group, or any controlling private power.[6]

When in 1934 General Butler blew the whistle on a group of businessmen conspiring to obtain the backing of the army to overthrow President

Roosevelt, it became clear that not even American democracy was safe from private power. The conspirators were activated by Roosevelt's conviction that the New Deal would end the Great Depression by replacing the market's invisible hand with government benevolence. Roosevelt wrote later: "'The New Deal' implied that the Government itself was going to use affirmative action to bring about its avowed objectives rather than stand by and hope that the general economic laws would attain them...the American system visualized protection of the individual against the misuse of private economic power, the New Deal would insist on curbing such power."[7]

President Theodore Roosevelt also recognized the existence of this invisible government: "Behind the ostensible government sits enthroned an invisible government owing no allegiance and acknowledging no responsibility to the people. To destroy this invisible government, to befoul the unholy alliance between corrupt business and corrupt politics, is the first task of statesmanship today."[8]

The malign influence of business on governments has been documented with a legion of examples by many authors.[9] We will dwell on this issue only insofar as it impacts the ability of liberal democracy to deliver sustainable environmental outcomes.

The corporation is an institution with a structure and imperatives that direct the actions of those within it. But it is also a legal institution whose existence and capacity to operate depend upon the law. Its legally defined mandate is to pursue, relentlessly and without exception, its own self-interest regardless, of the often harmful consequences it might cause to others.[10] As a result the corporation has become like a heartworm, *Dirofilaria immitis,* eating the heart out of democracy.

"Profit above all else" is best illustrated by the involvement by corporations in the financing of Hitler's rise to power and his war effort, from 1939 to 1945, as researched by Antony Sutton:

Wall Street financed German cartels in the mid 1920s which in turn proceeded to bring Hitler to power...the financing for Hitler and his SS street thugs came in part from affiliates or subsidiaries of US firms, including Henry Ford in 1922, payments by IG Farben and General Electric in 1933, followed by Standard Oil of New Jersey and I.T.T. subsidiary payments to Heinrich Himmler up to 1944...US multi-nationals under the control of Wall Street profited handsomely from Hitler's military construction program in the 1930s and at least till 1942...these same international bankers used political influence in the US to cover up their wartime collaboration and to do this infiltrated the US control commission for Germany.[11]

There is no excuse that those concerned did not know what they were doing. Standard Oil was assisting the development of synthetic gasoline for the German war effort and, as a result, received written protests from the U.S. War Department. It seems that even in the early 1940s corporate America had the power to do just what it wanted to make a buck. Sutton has never been taken to court for his many books and disclosures about American corporatism.

There are thousands of examples of corporate behavior that is harmful to society. Conversely the case is often made detailing the benefits to society of discoveries and innovations that lead to advances in society. It is not for us to weigh the evidence, our role is to see how corporatism must be controlled if the environmental crisis is to be averted.

Today, in the sphere of health, there is ample evidence of practices that harm society. The pharmaceutical industry has an embracive grip on U.S. government policy, research, and medical education.[12] The same is true in most other countries. These examples involve the health and lives of individuals and are well documented. Particularly well documented from legal class actions are the activities of over half a century of the tobacco industry. Despite their exposure the modus operandi is little changed, and recently it was reported that the head of British American Tobacco gained personal access to Mr. Blair to persuade him not to hold an enquiry into allegations that the firm was colluding with criminals to expand the use of its product via the black market.[13]

The tobacco industry has worked by providing funds to a large number of organizations with a scientific or community front. They operate by selecting an occasional scientific article that suggests that smoking may not cause lung cancer or that passive smoking is not injurious. This article is then used to undermine the majority scientific opinion. Public relations companies and compliant scientists work both privately and publicly to muddy the scientific waters of causation. The creation of doubt was the thrust of this activity. George Monbiot in his book *Heat*[14] describes the fake citizens group The Advancement of Sound Science Coalition (TASSC), created by a public relations firm at the behest of a tobacco manufacturer to muddy the waters on passive smoking. It was necessary to create the impression of a grassroots movement—one that had been formed spontaneously by concerned citizens to fight overregulation. It should portray the danger of tobacco smoke as just one unfounded fear among others, such as concerns about pesticides and cell phones. TASSC was to be a national coalition to educate media, public servants, and the public about junk science.

The research by Monbiot indicates that TASSC also receives funding from Exxon and finances www.junkscience.com, which propagates climate change denial. This would seem to be a remarkable coincidence of disparate causes, but the refined methods to debunk passive smoking can be used to rubbish other science inconvenient to industry. Indeed, as detailed by

Monbiot, an employee of TASSC writes for the Junk Science column on the Fox News web site, which debunks the dangers of passive smoking and climate change, without indicating the relationship to TASSC.

The promotion of tobacco continues to cause ill health and misery to millions of individuals. Climate change is also causing ill health to an increasing number of individuals, and deaths are being documented by the World Health Organization. It is an indictment of governance systems that those activities of the world's most profitable corporations can continue with impunity. Perhaps more relevant to our discussion is the psychological makeup of those who participate. They deal in death and destruction for profit, the endpoint of liberty conveyed by democracy upon corporate empires. It is impossible to grasp the psychological mechanisms by which this is rationalized. One has to turn to the philosophical thoughts of John Gray that it is just part of our nature. He states that "the destruction of the natural world is not the result of global capitalism, industrialization, 'Western civilization' or any flaw in human institutions. It is the consequence of the evolutionary success of an exceptionally rapacious primate. Throughout all history and prehistory, human advance has coincided with ecological devastation."[15] If we accept this view we can regard capitalism and democracy as efficient facilitators for our continually expanding needs. There is a message to change the system.

The influence of corrupt business referred to by Theodore Roosevelt is enacted today by political donation. In many liberal democracies, large sums are donated privately to political parties for election campaigns. In the United States of America this amounts to about 90 percent of the millions of dollars spent.[16] This buys access and influence. In Australia the situation is not much better, and in both countries it is possible to see this influence on fossil fuel and renewable energy policies.

Groups such as the Bilderberg Group, comprising the heads of European and American corporations, political leaders, and intellectuals, meet the description of an invisible government. With no reporters allowed, this group of big business leaders, including David Rockefeller and Timothy Geithner (President of the Federal Reserve Bank of New York), meet behind closed doors to steer international policy. No one is permitted to publicly discuss the contents of these meetings, where the fate of humanity is decided.

Buying influence by the corporate empires is not limited to political donations and pressure from big business think tanks. The largest conservation groups now accept considerable funding from corporate partners and governments, for example the U.S. Agency for International Development. Conservation groups have had a policy to attract these funds, and business has shown no reluctance to oblige. As a result of these links there is evidence of reluctance by conservation groups to criticize the environmental record of donors.[17] Such funding may seem to be necessary because of meager income from the public. But why does the public fail to donate? It may well be that massive corporate funding and government assurances enable the public to think that all is well with the environment. The influence and control of Theodore Roosevelt's "invisible government" now extends throughout society.

Each nation is under enormous pressure to either participate in globalization or face the overwhelming threat of destitution. The fundamental thrust of globalization has been the creation of an intermeshing global financial system that weaves together the economically successful liberal democracies and an increasing number of countries that are not democratic. Economies are freed to compete for investment that will create jobs and so preserve the fortunes of governments. Competition to produce cheaper goods and the free movement of labor have reduced the status of the employee by denying security and respect. Worker protection and health have been eroded, and families stressed by the need for more than one job. Nations are not in charge of their own destiny.

Today, this dilemma is illustrated in Germany, which, in its period of unprecedented economic growth from 1950 to 1970, developed a comprehensive welfare state. Today, private capital has deserted Germany. The rights and costs of labor have to be reduced to enhance market performance. To be competitive Germany must reduce its social provisions and create low-paid U.S.-style jobs and so follow the path of market reform already undertaken by its competitors. In the words of Alan Freeman, "[Chancellor] Schroder's reforms are a delusionary reversal of the relation between economy and society. Justice and democracy are sacrificed on the altar of a mythical market presented, like the Gods of an ancient religion, as a force outside society instead of a creation of it. They are a political choice dressed up as a market necessity."[18] In reality there is no end to the demands of the market. The calls for reduced corporate taxes and government spending continue as competition intensifies. This is why the liberal democracies cannot and will not address the environmental crisis; it is not possible within the economic system to which they are wedded. It cannot be funded.

The market economy controls the populace by the cult of consumerism. De Botton in *Status Anxiety*,[19] explains that democracy unleashed an expectation of ever-growing personal prosperity. Even if inequality remained, the option of success was always there like a lottery ticket. The mythology of the system, as illustrated by television shows such as *The Apprentice*, is that with hard work and good luck, we can all be rich. Woe to anyone who doubts this, for to do so is to challenge the system. In 1835 Alexis de Tocqueville wrote in *Democracy in America* that "Americans were often restless in the midst of their prosperity" and, in visiting the United States, he found that affluence did not stop Americans "wanting ever more and from suffering whenever they saw someone else with assets they lacked."[20] This avarice has grown during two centuries fuelled by the media depicting the lives of those with higher status and by the psychologically based advertising of goods for every perceived need. There was no going back to the hierarchical system, which, as de Botton reminds us, "despite all its wretchedness enjoyed several types of happiness which are difficult to appreciate today...one found inequality in society but men's souls were not degraded thereby."[21] Thus, in the eighteenth century there was freedom for those on the lowest rungs of society not to take the achievements of others as a reference point and find themselves wanting in status. By contrast liberal democracy has riddled society with status anxiety that cannot be assuaged. The rich feel that they cannot afford all they need,[22] and consumerism is married to depression, debt, and dissatisfaction. Yet consumerism has become the engine of capitalist society, consuming the earth's limited resources and creating jobs to stoke it. This fundamental nature of democracy is unsustainable.

THE ROLE OF THE MILITARY INDUSTRIAL COMPLEX

The manufacture of weapons has become one of the biggest economic growth industries in the world. Figures for military budgets in 2000 show that North America and Western Europe have 65 percent of the global military budget and, of the 14 biggest budgets, 9 are in the liberal European democracies. The United States and its allies have 75 percent of the global military budget; by comparison China has 3 percent. Incredibly, about half of the discretionary budget of the United States is military; by comparison, 2 percent is spent on environmental protection.[23] The five permanent members of the UN Security Council are the five biggest arms exporters in the world. Over five years British arms sales to countries in Africa increased from 52 million pounds to 200 million. Many of these countries are poor and have failing economies. In commenting on this situation, the *British Medical Journal* identifies these arms sales with the fuelling of conflict and massive health problems. The exporting countries place their short-term interests above those of the international community.[24]

Arms production is entrenched in the economic success of many Western democratic countries, and its cessation would induce an economic withdrawal convulsion followed by economic depression. This is particularly so in the United States, where arms spending is half of the discretionary budget. Weapons of war are promoted and sanitized as "security systems," and it is promoted by the arms industry that every nation, even when ravaged by poverty, has to have them for security against adjacent countries. A study of the arms industry illustrates the influence of corporate manufacturers on democratic governments. As with the oil industry, there are intimate links between the two partners, including movement of senior players between industry and government that ensures serious conflict of interests.

Let us consider one recent example. The potential covert role of this industrial complex in the creation of war has to be questioned in view of the Iraq War that was initiated on false pretences. We the public were told that excellent, completely reliable intelligence indicated that Saddam Hussein had weapons of mass destruction and that there were clear intelligence links between Iraq and al-Qaeda. In October 2004, the U.S. CIA admitted that there was at the time no such evidence. Secret UK high-level confidential briefing papers said that there was no legal justification for the war and that even if representative government was established in Iraq, they would still seek to obtain weapons of mass destruction. The British Spy Agency MI6 thought that Saddam did not have biological agent production facilities and only small stockpiles of chemical weapons. Peter Ricketts, the policy director to the then British Foreign Secretary Jack Straw, said to the British Prime Minister Tony Blair that the United States' attempt to link Iraq and al-Qaeda was unconvincing: "It sounds like a grudge match between Bush and Saddam." Straw also felt that there was "no creditable evidence to link Iraq to Osama bin Laden and al-Qaeda."[25] Of several possible reasons why the U.S. president perpetrated the war, oil and military needs were very important. The building of massive bases in Iraq indicates that the United States will protect its oil interests there despite an unwinnable war. The dispatch of more arms to the oil-producing regions around the world is food for the military and the arms manufacturers.

The Iraq war also illustrated the deceit and denial of truth in public affairs discussed in the next chapter as a failure of liberalism. Despite inadequate intelligence the president insisted that weapons of mass destruction were present in Iraq. As a result, the results of a survey by the Program on International Policy Attitudes (PIPA) showed that 72 percent of Bush supporters believed that before the war Iraq had weapons of mass destruction. In reality the weapons inspector Charles Duelfer concluded that "Saddam Hussein did not produce or possess any weapons of mass destruction for more than a decade before the U.S. led invasion," and that UN inspections had "curbed his ability to build or develop weapons."[26] Even after the publication of these conclusions, 57 percent of Bush supporters believed that the Charles Duelfer report concluded exactly the opposite—namely that Iraq either had weapons of mass destruction or a major program for developing them. The same study also found that 75 percent of Bush supporters believed that Iraq was providing substantial support to al-Qaeda, and 55 percent believed that this was the conclusion of the 9-11 Commission.[27] In reality, again, the 9-11 Commission concluded that there was no "collaborative relationship" between al-Qaeda and Iraq.[28] This is an example of reality denial, which has become all too common in public life today.

This discord between reality and the views of Bush supporters is both striking and alarming. We suggest that such results will not be restricted to Bush supporters, but would apply to a wide number of public issues across a broad spectrum of the public. This is the "mobocracy" feature of liberal democracy, the uncanny ability of the elites to fool most of the people, most of the time, on most issues.

William Blum, in two important books, *Rogue State* and *Killing Hope,*[29] has documented how the United States has attacked and subverted 23 nations that have not directly threatened the United States, primarily to serve the interests of transnational corporations. These nations include China, 1945–46, Korea, 1950–53, China, 1950–53, Guatemala, 1954, Indonesia, 1958, Cuba, 1959–60, Guatemala, 1954, Indonesia, 1958, Congo, 1964, Peru, 1965, Laos, 1964–73, Vietnam, 1961–73, Cambodia, 1969–70, Guatemala, 1967–69, Granada, 1983, Libya, 1986, El Salvador, 1980s, Nicaragua, 1980s, Panama, 1989, Iraq, 1991–99, Sudan, 1998, Afghanistan, 1998, and Yugoslavia, 1999. Of course most of these regimes were oppressive and places where we would not want to live. But U.S. actions have resulted in the loss of millions of lives—all this from a liberal democracy!

Supporters of liberal democracy often argue that one of the virtues of the system is that it enables internal conflict over who should govern to be settled without violence. Those who argue in this fashion seldom consider the objection that this lack of violence only applies to internal relations within the liberal democracy. As Blum has shown, the record of the United States as a leading exemplar of liberal democracy is appalling on the question of external wars and aggression. The United States has killed more people—rightly or wrongly—in these wars than Nazi Germany ever did (which of course is not to legitimize that regime). As a liberal democracy it is built upon the foundations of the blood and suffering of others.

The first part of our argument for the failure of liberal democracy is complete. Liberal democracy has essentially never been tried because society has been ruled by elites who set the social agenda and manipulate both the economy and public opinion through their control of money and the media. This does not necessarily have to be at the relatively crude level of brain washing, but it typically involves giving a selective and biased slant to issues or outrightly limiting information so that alternative viewpoints on a range of environmental and social issues are not publicly discussed. In particular, debates such as immigration and population issues are given a clear economic slant rather than an ecological one, so that important issues are never discussed in their proper ecological context.

To this one may say all the more reason for a democratic revolution to take away the power of the elites. But there are good reasons for doubting whether this can occur. There is, as Somit and Peterson have argued, a biological basis to authoritarianism.[30] We will now argue that even if the elites in capitalist society could be controlled, there remain good theoretical reasons for doubting whether democracies could ever work even in principle.

THE BIOLOGICAL BASIS OF AUTHORITARIANISM

Albert Somit and Steven Peterson in their book *Darwinism, Dominance, and Democracy*[31] have attempted to address why it is that throughout human history, the vast majority of political societies have seen the dominance of an elite over a submissive majority. Authoritarian regimes are the norm, while democracies are in a minority. This is certainly true of the past, but also true today in what democratic theorists like to call the "age of democracy." Somit and Peterson note that "authoritarian regimes are notable by their presence and persistence, democracies by their infrequencies and impermanence."[32]

The United States in the mid-nineteenth century was the first of the modern democracies, the first democracy since the Roman Republic 2,000 years previously. In the late twentieth century the democracy theorist Tatu Vanhanen lists the number of democracies as 61 or 41 percent in a total of 147 countries.[33] This study utilizes a definition of democracy based upon a majority rule system of representative government and does not include what some theorists believe is an equally important characteristic of democracy, the rule of law. The rule of law is the idea that the legal system should be based upon an independent judiciary and where all people are under the law, with political and civil liberties protected against arbitrary government infringement in a "lawful" society. Robert Dahl sees the number of countries that are democracies as having peaked in the 1950s and now being in decline.[34] Somit and Peterson conclude that many so-called democracies in Asia and South America drop out of the list (e.g., so-called illiberal democracies) when such factors such as censorship of the media, electoral fraud, corruption of government, and human rights violations are considered. Excluding microstates with a population of less than one million people there are 29 democratic macronations. Although doubts could be raised about some nations on the list, such as Ecuador and possibly Israel given its human rights violations, on most scholarly accounts democracies constitute a minority of the world's existing governments—and that number is falling. Freedom House, the organization that examines the degree of liberty of nations, documents a fall in the percentage of the world population living under free systems (i.e., liberal democracies) from 36 percent as of January 1981 to 20 percent in January 1996.[35]

Contrary to political theorists such as Francis Fukuyama, liberal democracy as an ideological system is not dominating the earth, although the extraordinary influence of one so-called liberal democracy, the United States, may give that impression. The numbers of democratic countries increased for a time in the past century and then diminished. Most of the newly created democracies of the twentieth century did not survive. There is no evidence that democratic forms of government are more stable than nondemocratic forms. A study by T.R. Gurr of the life of polities found that the average polity lasts only 32 years.[36] Like organisms, political systems come and then go, are born and then die.

Democracies then are not the natural state of humanity by any statistical measure. There are prerequisites for democracy, and the essential ingredients have been debated long and hard.[37] Factors include a generally equitable distribution of wealth, a relatively high degree of literacy and education (i.e., to be able to vote meaningfully),[38] a comparatively high energy consumption per person, and generally a homogeneous ethnic basis to society in the sense of there being, despite multiple minorities, a dominant ethnic and linguistic group (although, as illustrated by the Canadian province of Quebec, there may be two such groups in constant tension if one group desires to break away).[39] With the notable exception of Swiss democracy, democracy is less likely to *evolve* in multiethnic, multireligious states that normally require an authoritarian structure to hold such states together (e.g., the former Soviet Union and Yugoslavia).[40] Many important democracies are democracies by descent: The United States, Canada, and Australia deriving their democratic structure from Great Britain. Colonialism has been an important force in spreading democracy. Nevertheless democracy failed to take root in most former colonies, especially those of Africa.

It is not our aim to define any sort of formula by which democracies may arise, if there is such a formula. It is possible that democracy in both the modern and ancient world was a unique event like the emergence of a particular species of animal. Perhaps a better analogy is that the emergence of democracy is like the existence of phenomena in society like celibacy, phenomena contrary to our biological nature but which can be maintained through specific cultural structures and interventions. Somit and Peterson, who adopt a neo Darwinian evolutionary approach to the understanding of human behavior, believe that there is "an innate predisposition to dominance relations between individuals and for hierarchical social structures characterized by distinct differences of rank and status."[41] We will discuss this further.

THE NATURAL STATE OF AUTHORITARIANISM

Why then is the authoritarian state a natural choice for humanity? It is not necessarily a choice, it happens, because, as Richard Dawkins wrote, "If you wish to build a society in which individuals cooperate generously and unselfishly towards a common good you can expect little help from biological nature."[42] When Rousseau said that man was born free, this was far from the truth. We may not be happy with the thought, but there is much evidence to indicate that our evolutionary past dictates our instinct and behavior.

On reviewing the scientific evidence to substantiate this, Robert Winston concludes that while people have no problem accepting our evolution from some form of ape, few of us accept the psychological implications. "*Homo sapiens* not only looks, moves and breathes like an ape, he also thinks like one. Not only do we have a Stone Age body, with many vestiges of our past, but we also have a Stone Age mind."[43] This mind is ruled by such basic instincts of fear and flight by which automatic physiological responses occur in threatening situations, and by the primacy of the sexual instinct to ensure survival of the species. The latter is the main determinant of our quest for power, goods, and status, and when the chips are down, is more important to us than the governance system that we use to obtain it.

The modular theory of evolutionary psychology suggests that humans are born with minds that contain complex psychological mechanisms or modules so that the brain is hardwired for a wide range of behaviors and instincts that are shared by all humanity. These range from an inherent fear of snakes to an innate structure of the brain that allows us to learn language—according to the work of Chomsky.[44] The modular theory is supported by studies on patients who have injury to the brain localized by brain scanning, which shows a range of disabilities in speech and recall of words. These functions cannot be learned to any significant degree by undamaged parts of the brain. This is not an agreeable theory for humanity to accept, for it offers little hope for reform! Indeed other scientists believe that there is much plasticity in the brain that is adapted by our experience of the world around us. As with all diametrically opposed theories in science, the truth will encompass some of both theories with the modular theory preeminent.

With the modular theory in mind, it is important to note that Somit and Peterson believe that our social evolution in tribal systems is framed around "dominance and submission, command and obedience."[45] Dominance is a relationship between different individuals that is usually established by threat and display. It serves the important role of preventing disputes that might lead to injury and turmoil. In evolutionary terms, violence would not be good for reproductive success. This system is seen in primates where it contributes to reproductive success, and a hierarchy is established that leads to social stability. Humanity uses dominance and submission to organize society. The reproductive intent is more hidden in the cloak of power and prestige of those who are leaders either elected or appointed.

Within democracy we are always on the move towards authoritarianism. Political parties are hierarchical. Often they have cabals, each of which has its own hierarchy that selects its candidates for government. We have to have visible and directive leaders, even though we may recognize that the leader is constructed from cardboard and painted by spin-doctors and advertisers. Government, opposition and corporatism is hierarchical and cannot be challenged from within without potential injury. An exposure of misdoing or corruption by a whistle-blower is not accepted as a service to society. Instead of gratitude, there is discomfort, "outing," and unemployment. Those elected to leadership by democracy often move to authoritarianism by using the system to retain power or to wage war. In particular they consort with the rich and powerful corporations to usurp the needs of society, even to the extent of destroying other democracies if they fail to satisfy the mould sought by corporatism, for example, Allende's Chile. All these human traits are genetic barriers to the sustainability of democracy.

Whatever social structure is freely created, it inherently becomes hierarchical and authoritarian. It is difficult to comprehend how a simple universal message of love and humility espoused by Christ and the disciples could be transformed into the pomp, power, and authoritarian dogma of the Roman Catholic Church.

Obedience is part of this hierarchical system, and disobedience is rare. This is also an impediment to democracy. Obedience is expected within a so-called democratic party where the members are kept in line by whips, and in the workplace where questioning of roles can be insubordination. An order may be accepted when it involves personal sacrifice, and orders that are morally reprehensible such as torture, massacre, and genocide are often carried out with alacrity by individuals, formerly good family stalwarts of society. Obedience is necessary for the functioning of the killing machines, the armies trained by democracies as well as the tyrants. The scientific study of obedience using electric shocks shows that individuals have an ingrained ability to obey even when injury is conferred on others.[46]

Observation of our closest primate relatives, chimpanzees, reveals a social and hierarchical structure uncannily similar to our own. Their society functions with a hierarchy based on dominance and submission. The dominant male is the leader because of strength and creation of alliances. Murder and organized violence are part of their society just as they are in ours. For example, male chimpanzees form alliances to seek revenge when a friend is killed. War parties are formed from mature males who have grown up together, and the anticipation of battle may produce sickness and vomiting through fear. These activities closely resemble the male bonding and platoon formation in human wars. This common behavior is summarized by Potts and Short as follows: "The unique and bloody common characteristic of the chimpanzee *Pan troglodytes* and *Homo sapiens* is a propensity for a close knit group of mature males to drop what they are doing, venture stealthily and deliberately into the territory of a neighboring group, seek out one or more individuals they can outnumber, and then beat the living daylights out of them. This behavior has not been found in any other animal and it has all the attributes of a war."[47] Indeed, both societies sometimes choose warfare as a strategy, even perhaps to the extent of preemptive strikes. Both societies can revel in the sight of violence, one need look no further than the television schedules. Liberal democracy provides but sheep's clothing for its selfish authoritarian genes.

NOTES

1. Paul Sheehan, *The Electronic Whorehouse* (Pan Macmillan, Sydney, 2003).

2. Le Carre, cited by Sheehan. Ibid., p. 34.

3. Stern Review, "The Economics of Climate Change," at <http://www.hm-treasury.gov.uk/independent_reviews/stern_review_economics_climate_change/stern_review_report.cfm>.

4. Milton Friedman, quoted in interview, Joel Balkan, *The Corporation, The Pathological Pursuit of Power and Profit* (Constable & Robinson Ltd, London, 2004), p. 34.

5. Arundhati Roy, Sydney Peace Prize Lecture, "Peace and the New Corporate Liberation Theology," Delivered November 3 2004 at the Seymour Centre Sydney .

6. Samuel Rosenman (ed.), *The Public Papers and Addresses of Franklin D. Roosevelt, Volume Two: The Year of Crisis, 1933* (New York, Random House, 1938), as cited by Cass Sunstein, *The Partial Constitution* (Harvard University Press, Cambridge, MA, 1993), pp. 57–58.

7. Ibid.

8. <http://en.wikiquote.org/wiki/Theodore_Roosevelt>.

9. Balkan, *The Corporation,* from note 4.

10. Balkan, *The Corporation,* from note 4.

11. Antony Sutton, *Wall Street and the Rise of Hitler,* at <http://reformed-theology.org/html/books/wall_street/index.html>.

12. Marcia Angell, *The Truth About Drug Companies: How They Deceive Us and What to Do About It* (Random House, New York, 2004).

13. "Tobacco Giant Gained Secret Access to Blair," *The Guardian Weekly,* November 5, 2004.

14. G. Monbiot, *Heat: How to Stop the Planet Burning* (Allen Lane, London, 2006).

15. John Gray, *Straw Dogs: Thoughts on Humans and Other Animals* (Granta Books, London, 2002), p. 7.

16. Sally Young, *The Persuaders. Inside the Hidden Machine of Political Advertising* (Pluto Press, Sydney, 2004).

17. M. Chapin, "A Challenge to Conservationists,"*World Watch Magazine,* November/December 2004.

18. Alan Freeman, "Why Not Eat Children?" *The Guardian Weekly,* October 22, 2004.

19. A. de Botton, *Status Anxiety* (Pantheon Books, New York, 2004).

20. Alexis de Tocqueville, quoted from de Botton, Ibid., p. 52.

21. Ibid., p. 52.

22. Clive Hamilton, *The Politics of Affluence* (The Australia Institute, Canberra, December 2002).

23. Ron Nielsen, *The Little Green Handbook: A Guide to Critical Global Trends* (Scribe Publications, Melbourne, 2005).

24. Editorial, "Arms Sales, Health and Security," *British Medical Journal,* vol. 326, 2003, pp. 459–460.

25. *Los Angeles Times,* April 11, 2002, at <http://www.cooperativeresearch.org/entity.jsp?entity=jack_straw>.

26. "Bush Supporters Misled," October 2004, at <http://www.misleader.org/daily_mislead/read.html?fn=df10222004.html>.

27. Ibid.

28. Ibid.

29. William Blum, *Rogue State* (Zed Books, London, 2002); William Blum, *Killing Hope* (Zed Books, London, 2003).

30. A. Somit and S.A. Peterson, *Darwinism, Dominance, and Democracy* (Praeger, Westport, London, 1997). We are in debt to their bibliography for the references in notes 33–38.

31. Ibid.

32. Ibid., p. 35.

33. T. Vanhanen, *The Process of Democratization* (Crane Russak, New York, 1990).

34. R. Dahl, *Modern Political Analysis* (Prentice Hall, Englewood Cliffs, NJ, 1991).

35. Somit and Peterson, *Darwinism, Dominance and Democracy,* from note 30, p. 13; A. Karatnycky, "Democracy and Despotism: Bipolarism Renewed?" *Freedom Review,* vol. 27, January–February 1996, pp. 1–16; S. Huntington, "How Countries Democratize," *Political Science Quarterly,* vol. 106, 1991–1992, pp. 599–616.

36. T.R. Gurr, "Persistence and Change in Political Systems, 1800–1971," *American Political Science Review,* vol. 68, 1974, pp. 1482–1504.

37. F.W. Miller and M.A. Seligson, " Civic Culture and Democracy," *American Political Science Review,* vol. 88, 1994, pp. 635–652.

38. T. Vanhanen, *The Emergence of Democracy* (The Finnish Society of Sciences and Letters, Helsinki, 1984).

39. R.E. Burkhart and M.S. Lewis-Beck, "Comparative Democracy: The Economic Development Thesis," *American Political Science Review,* vol. 88, 1994, pp. 903–910.

40. J.W. Smith, et al., *Global Meltdown* (Praeger, Westport, CT, 1998).

41. Somit and Peterson, *Darwinism, Dominance and Democracy,* from note 30, p. 51.

42. Richard Dawkins, *The Selfish Gene* (Oxford University Press, Oxford and New York, 1971), p. 59.

43. Noam Chomsky, cited from Robert Winston, *Human Instinct: How Our Primeval Impulses Shape Our Modern Lives* (Bantam Books, New York, 2003), p. 17.

44. Winston, *Human Instinct,* p. 93.

45. Somit and Peterson, *Darwinism, Dominance and Democracy,* from note 30, p. 51.

46. Stanley Milgram, *Obedience to Authority: An Experimental View* (Taristock, London, 1974).

47. Malcolm Potts and Roger Short, *Ever Since Adam and Eve: The Evolution of Human Sexuality* (Cambridge University Press, Cambridge, 1999), p. 194.

— 7 —

Moribund Liberalism: The Collapse of Liberal Values

The tone and tendency of liberalism...is to attack the institutions of the country under the name of reform and to make war on the manners and customs of the people under the pretext of progress.
—*Benjamin Disraeli, 1882*

THE IRON CAGE OF LIBERALISM

So far, we have criticized democracy and liberalism, arguing that not only are they incapable of providing an adequate response to the environmental crisis, but that in many cases they have acted as an accelerant, adding fuel to the wildfire savaging the ecology of the planet. We have also pointed out that although as a matter of logic and definition liberalism and democracy can be conceptually distinguished (so that there can be and are, illiberal democracies), in the West, liberalism and democracy have been organically linked. Perhaps this was so to some degree even in ancient Greece, as we see when we reflect upon Plato's criticism of the institution of democracy in the *Republic*.[1] There, Plato noted that democracy rests upon the desire for liberty, which ultimately generates diversity and fragmentation. People demand the right to live as they see fit, so that ways of living proliferate and violently clash with one another. After a time this conflict becomes too great, leaving society—or what remains of it—to be held together by a despot.

Following Plato's lead we will also argue in this chapter and the next that there are inherent contradictions in the notion of liberal democracy that ultimately doom such a political system to extinction. Some have seen liberal democracy as the end state of human history. In 1989 the conservative American policy theorist Francis Fukuyama published an article entitled "The End of History?" in the journal *The National Interest*.[2] He argued that liberal democracy is the "endpoint of mankind's ideological evolution" and thus the "final form of human government." It is one of our aims to question this claim by showing that liberal democracy is essentially a moment in human history and that its time will pass, at least on the present "business as usual" scenario, unless there are radical reforms as discussed in chapter 10.

The present chapter moves away from the previous argument of this book, which was that liberal democracy leads to a tragedy of the commons and hence to environmental destruction. Here our intent is to summarize criticisms that have been made of liberal democracy by philosophers and policy theorists who do not generally write on environmental matters, and who may hold views on such questions that are radically different from ours. Nevertheless, these writers are united in seeing fundamental problems in the liberal democratic world view. We will show that liberal democracy is also self-undermining in the social sphere as well. The criticism of democracy, therefore, is not confined to environmental issues; rather there is a consensus of concern from those working in other spheres. Our argument is that if liberalism and liberal democracy cannot solve small problems such as, say, the conflicts associated with multiculturalism, how can it possibly cope with grand-scale, civilization-threatening problems such as the environmental crisis? If liberalism is flawed as a system of ideas then it is logical to suppose that it cannot meet such a challenge.

Our critique also differs from the ecological critique of liberalism made by a vast number of writers in the last 40 years. Writers loosely described as "deep ecologists" or "dark greens" have argued that the Western philosophical tradition, as epitomized by liberalism, but not limited by it, is flawed because it has led to a destructive unsustainable ethos that gives only instrumental or economic value to nature. This is the critique of modernity made by a diverse number of thinkers.[3] For example, Arne Naess[4] distinguished between shallow and deep ecology, with deep ecology giving intrinsic value to nature and shallow ecology bestowing only instrumental value. Ethics has been primarily human centered, seeing moral value arising from human beings, and liberalism is a classic example of that. Deep ecology sees human beings as only one valuable species among many, and humans, despite their technological sophistication, are not more important than nature. Liberalism then is flawed for being a "human chauvinist" moral theory, a theory with an unjustified bias in favor of humans. We agree that this is a fundamental problem of liberalism, but unlike these authors we will detail how liberalism is destructive in other human and social spheres. We will see that the liberal attitudes that have corrupted the concept of environmental sustainability, for this is incompatible with the growth economy, are the same ones that conflict with human values.

IS LIBERALISM A SOCIALLY DESTRUCTIVE IDEOLOGY?

This claim that liberalism is a socially destructive ideology has been made in the recent past by conservative thinkers. Malcolm Muggeridge, a Christian conservative thinker, saw liberals as possessing a death wish. Reflecting on the cultural revolution of the West in the 1960s, he saw liberals as having a tendency to grovel before any tyrant and their regime, however brutal (Pol Pot, Mao, etc.), so long as the tyrant mouthed the appropriate platitudes about the "brotherhood of man."[5] Seeing humans as fallen and cursed by original sin, Muggeridge saw the liberal as possessed by an irrational necessity to abolish a degenerating culture and reconstruct it to conform to his or her own prejudices, revolving around the idea that human beings are fundamentally good.

The notion that liberalism is grounded upon a fundamentally mistaken philosophy of human nature is also expressed by the conservative, former communist, writer James Burnham.[6] For Burnham, liberalism is an ideology of national suicide that ultimately erodes a nation's will to survive. During the cold war French writer Jean-Francois Revel thought that communism would ultimately defeat liberal democracy.[7] Liberal democracy may only have been an historical accident. This political system always allows internal enemies who seek to destroy the system to flourish (such as communists for Revel or environmentalists for corporate empires). Liberal democracies will self-destruct by following their own logic to extremes.

The conservative intellectual Paul Gottfried has agreed that, unlike nineteenth-century liberalism, the contemporary liberal state is concerned with promoting uniformity, not individuality.[8] For Gottfried liberalism now only pays shallow lip service to the philosophy of liberty; today the nanny state is more concerned with democratic socialization and social control. There is no real mobilization by the oppressed against the new class elites who run the state machine.[9]

Psychological weapons, with fear the most potent, are refined to maintain social control, power, and community silence. Fear allows those in power to enact sweeping counterterror legislation, spy on its citizens, kidnap, and torture in the name of their protection.[10] Secrecy and deception become a normal part of liberal democracy, as was the case in totalitarian communism. The tenets of liberalism, such as justice, are cast aside in the interests of political designs. Political attacks on the judiciary became more and more open in Western democracy, reflecting liberalism's propensity to cast aside the collective good in favor of individual liberty. In a speech just after her retirement from the U.S. Supreme Court, Sandra Day O'Connor took aim at those leaders whose repeated denunciation of courts for alleged liberal bias could be contributing to a climate of bias against judges.[11] The leaders were political right-wingers flourishing in the free-for-all of liberalism. History tells us that attacks on the judiciary are often the forerunner of dictatorship. Thus the masses become apathetic and lose hope, if they ever had any, of self-government and are pacified by sexual bread and drug circuses, a quiet tyranny of tits, TV, and consumer consolation. In this sense, George Orwell's 1984 has already arrived. Then the masses were pacified by Victory gin, "films football, beer and above all gambling filled up the horizon of their minds," and "there were some millions of proles for whom the lottery was the principal if not the only reason for remaining alive."[12]

Gottfried's line of thought is that it was liberalism that destroyed the old monarchic order and concentrated power in the emerging modern state of Europe. At the time of the 1917 Revolution, communists in the former Soviet Union found a concentration of power ready for them to take over. Liberalism thus, ironically, laid the ground for the Soviet gulag.[13]

As discussed in chapter 6, liberal societies are far from liberal, in that the number of people killed by liberal democratic governments in the name of universal human emancipation from the time of the Enlightenment to the second Iraqi war is far greater than the number of people killed by communist regimes (thought by some authorities to be in excess of 150 million).[14] Often those liberal democracies strutting their freedom and opportunity for all have transgressed the rights of others in the name of their own self-interest. The accusations of Harold Pinter ring true in his Nobel Lecture:

The United States supported and in many cases engendered every right wing military dictatorship in the world after the end of the Second World War. I refer to Indonesia, Greece, Uruguay, Brazil, Paraguay, Haiti, Turkey, the Philippines, Guatemala, El Salvador and of course Chile. The horror the United States inflicted upon Chile in 1973 can never be purged and can never be forgiven. Hundreds of thousands of deaths took place throughout these countries.[15]

The United States is not alone in its actions for other liberal democracies, the UK, France, and others, have behaved similarly to protect their power and economic interests.

The American conservative philosopher John Kekes concludes that liberalism is inconsistent "because the realization of these liberal values would increase the evils liberals want to avoid and because the decrease of these evils depends on creating conditions contrary to the liberal values."[16] A good example of this paradox is the liberals' advocation of both antiracism and multiculturalism and also the right of free speech, a matter to be discussed. For these thinkers, liberalism, in short, saws off the branch that supports it.[17] These points can be developed by briefly considering some arguments made along these lines by Paul Gottfried.

Gottfried points out that liberalism, in embracing doctrines such as hard multiculturalism, has generated further internal contradictions. For example, on the face of it, the 1972 French Gayssot Law seems reasonable enough. The law forbids "provocation to discrimination, to violence, or to hatred against a person or groups of persons by reason of their origin,"[18] Fair enough. Also prohibited is "public defamation of a person or group of persons by reason of their origin or belonging or non-belonging to an ethnic body, nation, race or determined religion."[19] Again on the surface this seems reasonable. But although such laws have been used to put Holocaust deniers in their place, they have also been used against those criticizing various aspects of France's immigration policy. One would have thought that a liberal democratic society would encourage, not suppress scholarly examination of its basic legal institutions.[20]

French actress Bridget Bardot's criticism of Muslim migrants' mistreatment of animals, for example, fell under the French race hate legislation. She narrowly escaped two years in prison. In Germany the use of ancient Germanic runic symbols (the same type of symbols as seen in movies such as *The Lord of the Rings*) has been banned because a small minority of neo-Nazi groups decorated CD albums with them. Even the use of the Irish Celtic cross, a Celtic Christian symbol, has been banned for fear that it may have racist implications. Canada has banned controversial, yet prima facie scientific texts on race and behavior, such as by Canadian psychologist J. Philippe Rushton[21] and a critique of America's immigration program by Wayne Lutton and John Tanton.[22] Yet there is no ban placed upon many American black rap songs, which often contain clearly racist and violent lyrics often expressing desires to murder white people and rape white women. Such albums often express racist sentiments towards whites, or "crackers" or "rednecks," as white people are called. It may be thought that this is an understandable revolt of an oppressed group of people against an elite group of people. Yet most of these rappers are not ghetto youth but very rich black Americans who produce their music for a largely white youth market, not for oppressed and poor black minorities who could hardly afford these expensive CDs on their welfare checks. The black rap music is the white middle class kids' revolt against their parents who pay the bills.

In Australia, race hate legislation was even used against a humane and sensible liberal journalist, Phillip Adams, for his controversial, but arguably right, condemnation of Americans for their support of the war on Iraq. Adams had said no more than an American critic such as Michael Moore had said, but an American in Australia was offended by Adams' condemnation of Americans and took him to the Human Rights and Equal Opportunity Court through a race hate complaint. We are not saying that such legislation is wrong in spirit, but it does seem to be inconsistently applied, and, as the Adams' case shows, can have some nasty and unanticipated uses. In the future it could easily be used as a weapon of oppression to silence critics on a number of issues. As we see from the above examples, it is already being used to silence critics of immigration.

Although counterterror legislation itself has not yet been used to explicitly suppress environmental criticism by labeling environmentalists as extremists, the legal system of the modern state has adequate means of doing so. Defamation laws in common law countries such as Australia are much stricter than in the United States. Australia has a poor legal framework for defending free speech, with no constitutional protection as the United States has in its First Amendment. Defamation law arose in England as a way of protecting the reputation of noblemen from criticism and public exposure. Today defamation cases are big business, where offended parties typically seek hundreds of thousands, if not millions of dollars of damage. Corporations and business people, typically developers, have made use of "SLAPP suits" to silence environmental critics of projects. SLAPP suits are strategically planned litigation brought against protesters to silence criticism—strategic lawsuits against public participation. This strategy is to threaten action against people who often have no more assets than their house with massive damage claims unless they cease their protest and apologize. In Australia, legislation such as the *Trade Practices Act* of 1974 (Commonwealth), which was originally devised as a form of consumer protection to produce a climate of fair trading, has been used against various environmental protesters by certain business organizations. The idea is to show that the protesters are frustrating trade by the protest itself, and massive damages are often sought. As we have no wish for such litigation against us, and in some cases even mentioning cases in discussion has lead to further litigation—the reader requiring more details will need to pursue this matter on the Internet through the use of any Internet search engine with appropriate key words.

Should one turn a blind eye to such inconsistencies in the name of tolerance? Liberals do so today just as a previous generation of the Left whitewashed the horrors and genocide of the communist regimes. But it did not make such horrors go away.

Liberals lack a fundamental ability to be able to face up to the internal contradictions in their own position. As Brian Appleyard in *Understanding the Present* has said with some rhetorical flourish:

> It is, I believe, humanly impossible to be a liberal. Society may advocate liberal tolerance and open-mindedness, but nobody practices it. In fact, this is what preserves liberal society. For a complete personal acceptance of scientific-liberalism would reduce the society to passive, bestial anarchy. There would be no reason to do anything, no decisions worth making and certainly no point in defending one position as opposed to another.[23]

The liberal difficulty in facing up to uncomfortable realities is well illustrated by the debate about whether feminism and multiculturalism are compatible. Liberals support women's liberation and equality with men even though practical equality in the workplace is not delivered by them. This parroting of equality is reminiscent of *Animal Farm* and "some animals are more equal than others."[24] Liberals also support antiracism, nondiscriminatory immigration programs, and allowing diverse cultures to maintain their traditions. However, fundamentalist Islam is strongly antifeminist and highly patriarchal. If in principle there is no reason for immigration restrictions based upon culture and religion, there is no reason why a nation such as France should restrict building upon its already significant Muslim population. But what if this in turn led to a cultural and ethnic change leading to a radical demographic change? This would undermine women's rights? Thus feminism and multiculturalism, products of liberalism, are mutually incompatible.[25]

The typical liberal response to such questions is to slam the questioner with abuse, usually calling the questioner a racist or fascist. But that doesn't solve the problem. The messenger may be silenced but the question remains. Political correctness is essentially about not asking these types of uncomfortable questions. Clearly some differences are more "different" than others.

ANYTHING GOES? THE PHILOSOPHY OF LIBERALISM

Philosophically, liberalism as a political philosophy is based on individualism (that all societies are a collection of individuals) and value hedonism (that satisfaction of individual's happiness and pleasure is all that the moral good can be). Such value hedonism is subjective, not objective: The standard of rightness depends upon the individual and not upon some eternal standard. Liberals have tended to be relativistic about truth as well: There is no objective standard of truth by which individual choices can be judged. For that reason free speech should be granted, for who knows where the truth may lie? This was the classic liberal defense of free speech given by John Stewart Mill in *On Liberty*,[26] but which seems to have been forgotten by modern liberals.

Mill's *On Liberty* illustrates why liberalism and democracy have been organically linked together: Democracy allegedly provides a solution to the philosophical problems that liberalism creates from its own internal contradictions. We do not know what the truth about certain things is, so it is best that we take a vote on the matter because anyone's view is as good as anyone else's.[27]

Although they may deny it, the economic rationalist writers in our daily papers are essentially relativists in their approach to the environment. For them the world can only be seen through the vision of economics. This explains why the Stern Review—*The Economics of Climate Change*—had more impact than the voluminous scientific reports of the Intergovernmental Panel on Climate Change (IPCC). A cultural maladaptation as defined by Stephen Boyden is being perpetrated.[28] The world is not a social construction as the relativists thought, but an economic construction. Human beings are not primarily biological organisms (of a complex cultural kind) but are economic agents. Neo-classical economists believe that marriage and most other social institutions such as law are best considered in economic terms—that is in terms of utility satisfaction. Concerns about the biophysical processes such as global warming are discounted because there is effectively no world outside economic reality.[29]

For example, it was reported by Anthony Barnett in *The Observer*[30] that an e-mail sent to the press secretaries of all Republican congressmen advising them as to what to say on the environment in the lead up to the November 2004 election was to tell that all is rosy. Global warming has not been proved, the world's forests are "increasing not decreasing" and the "world's water is cleaner and reaching more people." The e-mail goes on to say that "links between air quality and asthma in children remain cloudy," and that the U.S. Environment Protection Agency is exaggerating in its claim that at least 40 percent of American streams, rivers, and lakes are so polluted that they can no longer be used for drinking, swimming, or fishing.[31] The memo's sources for these statements are from those close to the heart of corporatism—right-wing neocon think tanks and scientists who have been funded by the oil industry. Scientific reality is thrown out of the window in favor of economic ideology.

It may be said in reply that relativists, or some of them, are attempting to defend the oppressed through debunking rational science. But relativism (the view that there is no objective truth and that all views are true from their own perspective) does not necessarily have to defend the Left and the oppressed as we have just seen. As a further example, the late Paul Feyerbend, a radical liberal philosopher of science, propounded that the logical conclusion of the liberalism of J.S. Mills was that any theory was as good as any other, that is, that "anything goes."[32] He proudly accepted that truth is relative as a badge of honor. Feyerbend defended extremist positions such as creationism, voodoo, and Nazism, all as viable liberal traditions. If reality is nothing more than a negotiated social construction then surely there can be nothing wrong with these traditions. It is only a matter of accident that relativism defends Left values: It could just as easily be used to defend right-wing extremism. Holocaust denial and denial of black slavery are some of the foul fruits of a denial of objective truth and advocation of the idea that history is a mere social construction.

REQUIEM FOR LIBERALISM

A number of philosophers and social theorists have seen the liberal order as being at an end. British philosopher Alasdair MacIntyre in *After Virtue*[33] sees liberalism as an intrinsically flawed philosophy, for while pretending to be a master system of morality it is really only one moral system among a competing plurality of alternatives and cannot supply an objectively true justification of its own foundations. Liberalism is seen to beg the question of its own truth by assuming the primary value of its fundamental concept: freedom. MacIntyre concludes his work by seeing liberalism as not a genuine morality at all, in the sense of providing a moral worldview compared to the heroic societies of Homeric times. Liberalism fails to provide a philosophy of life.[34] If one has no philosophy of life then one cannot accept the value of nature. Perhaps the sense of this loss of a heroic view of life is what has made films such as *Gladiator, Lord of the Rings,* and *Troy* so popular. MacIntyre sees liberalism as leading to the ultimate end of this social order, which will inevitably break down or fall apart from a kind of moral entropy. Advocating a type of communitarian survivalism, MacIntyre believes that only small state-independent Benedictine-style communities will survive the coming dark age that liberalism is creating.

Writing long before McIntyre in 1936, Lawrence Dennis[35] saw capitalism and communism as both doomed because of ecological scarcity, as there is a limit to economic growth. He was right about communism. With regard to capitalism, he argued that capitalism is more than just the private (i.e., nonstate) ownership of the means of production. The essence of liberalism, Dennis and others have argued, is to give greater consideration to private property rights than to human life. Thus modern liberal capitalism requires a market expanding in geometrical progression for its successful operation. The physical limits to growth dooms capitalism: "Even the harshest critics of modern capitalism have never for a moment questioned its ability to go on growing indefinitely in geometrical progression."[36] Of course that statement was made in America in 1936, and since that time many have asked that very question. Dennis believed that liberal capitalism would grow like a cancer, producing environmental destruction in its wake. The system will inevitably destroy itself, to be replaced by a type of steady-state authoritarianism.

William Ophuls in *Ecology and the Politics of Scarcity*[37] is one of the few ecology writers to reject democracy and favor an authoritarian solution to the environmental crisis. In the second version of the book, the anti-democratic focus has unfortunately been revised.[38] Nevertheless in his most recent book, *Requiem for Modern Politics,*[39] he returns to the theme of the rejection of liberalism.

The thesis for Ophuls' *Requiem for Modern Politics* is that modern politics is at an end because the concepts and values of the Enlightenment of individualism, liberty, and materialism are no longer viable. He states:

> Modern civilization, in all of its aspects and everywhere on the planet, is plunging ever deeper into a multiplicity of crises that call into question its governing principles, practices and institutions. In this "crisis of crises," there is one that has yet to receive the attention it deserves: the impending failure of liberal polity, the modern system of politics founded on the tenets of classical liberalism and the rationalistic philosophy of the Enlightenment. Liberal polity is based on intrinsically self-destructive and potentially dangerous principles. It has already failed in its collectivist form and, contrary to the view of many, is now moribund in its individualist form as well... Thus the three main components of modern civilization—liberal polity, exploitative economy, and purposive rationality—are riddled with inner contradictions. Civilization is therefore collapsing. As a result, the latent totalitarianism of modern politics is likely to manifest itself with increasing force in the years to come. In short, without a major advance in civilization, we confront a political debacle.[40]

Economic growth and development are the modern liberal state's raison d'etre—but these phenomena are challenged by ecological scarcity, the idea that there are limits to growth. These are not the only self-destructive tendencies in modern liberalism, Ophuls argues. Liberalism tends to moral entropy (i.e., moral decay) with individual selfishness destroying civil society: "liberal policies destroy themselves by devouring their own moral capital, the fund of fossil virtue they have inherited from the pre-modern past."[41] This can be seen in various shapes and forms: the destruction of civil society by a globalized market system;[42] education, which has become a prescription for intellectual uniformity; the decay of reason; crime; violence; and family breakdown. In short, "America exemplifies the process of growing barbarization that is pushing us towards a Hobbesian future."[43] For Ophuls, the liberal order has no future. Liberalism is also at an end.

The end of liberal America could be sooner than we may think. General Tommy Franks, who led the U.S. military operation to liberate Iraq, says that if America is hit by a weapon of mass destruction that causes large casualties, the Constitution will be discarded and the United States will have a military form of government. In one interview he said that the result of a weapon of mass destruction hitting the United States would mean "the Western world, the free world, loses what it cherishes most, and that is freedom and liberty we've seen for a couple of hundred years in this grand experiment that we call democracy."[44] He continued that "it may be in the United States of America—that causes our population to question our own Constitution and begin to militarize our country in order to avoid a repeat of another mass, casualty-producing event. Which in fact, then begins to unravel the fabric of our Constitution."[45] In this context it is worth quoting the more recent remarks of the Australian cosmologist John O'Connor who has reminded us that civilizations typically collapse from within:

Kenneth Clark in a famous television documentary series, *Civilization,* warned us that societies, however complex and solid they may appear, are in fact quite fragile. For example the almost total eclipse of the Greco-Roman civilization in western Europe after 600 years of predominance shows that collapse can occur when a society becomes exhausted... when its people become so used to the rights, privileges and material prosperity endowed by their civilization that they no longer value them sufficiently to defend, maintain and build on them.[46]

Similar sentiments have been expressed about the survival of America by the respected social theorist Chalmers Johnson in *The Sorrows of Empire.*[47] This is a perspective different from the one expressed in this book; Johnson sees America as a new Roman Empire, but a more enlightened one. Nevertheless the expansion of the American empire has led to the "sorrows of empire," including America becoming a debtor nation, owing more money than it is ever likely to pay back. International finance has a death grip on the throat of the American economy. Running an empire was expensive for the Romans, and it is even more expensive for the Americans. The arrogance of empire blinds leaders to basic realities: A combination of imperial over-stretching, rigid economic institutions, and an inability to reform weakens empires leaving them fatally vulnerable in the face of disastrous wars, many of which the empires themselves invited. There is no reason to think that an American empire will not go the same way and for the same reasons. However given the global reach of the American empire, the fall of America will be much like a large comet striking the ocean. The death of America will mean the death of liberal democracy.[48]

Liberal democracy likewise suffers from these sorrows of empire. The system is, in short, corrosive of social capital, the cultural glue that holds society together.[49] Although theorists differ about how and to what extent this corrosion acts, it is clear that act it does. The difficulties, contradictions, and dilemmas of liberal democracy are so great that that its demise is inevitable. What then will replace it, and what *should* replace it? The remainder of this book will consider these questions.

NOTES

1. The argument that Plato's critique of democracy should be interpreted along these lines is given by E.R.V. Kuehnelt-Leddihn, "A Critique of Democracy," *The New Scholasticism*, vol. 20, July 1946, pp. 195–238.

2. F. Fukuyama, "The End of History?" *The National Interest*, vol. 11, summer 1989, pp. 3–11; and F. Fukuyama, *The End of History and the Last Man* (Free Press, New York, 1992).

3. John Passmore, *Man's Responsibility for Nature: Ecological Problems and Western Traditions* (Duckworth, London, 1974); James Lovelock, *Gaia: A New Look at Life on Earth* (Oxford University Press, Oxford, 1979); Aldo Leopold, *A Sandy County Almanac* (Oxford University Press, New York, 1966); Arne Naess, "The Shallow and the Deep, Long-Range Ecology Movement: A Summary," *Inquiry*, vol. 16, spring 1973, pp. 95–100.

4. Arne Naess, "The Shallow and the Deep, Long-Range Ecology Movement: A Summary," *Inquiry*, vol. 16, spring 1973, pp. 95–100.

5. Malcolm Muggeridge, *Things Past* (Collins, London, 1978), pp. 220–238. A similar complaint is made by T. Suric, *Against Democracy and Equality* (Peter Lang, New York, 1990).

6. James Burnham, *Suicide of the West: An Essay on the Meaning of Destiny of Liberalism* (Jonathan Cape, London, 1965). Along the same lines is A.M. Ludovici, *The Specious Origins of Liberalism: The Genesis of an Illusion* (Britons Publishing Company, London, 1967).

7. Jean-Francois Revel, *How Democracies Perish* (Weidenfeld and Nicolson, London, 1983).

8. Paul Gottfried, *After Liberalism: Mass Democracy in the Managerial State* (Princeton University Press, Princeton, NJ, 1999).

9. M.A. Glendon, *Rights Talk: The Impoverishment of Political Discourse* (The Free Press, New York, 1991).

10. J. Risen, *State of War: The Secret History of the CIA and the Bush Administration* (The Free Press, New York, 2006).

11. J. Borger, "Former Top Judge Says US Risks Edging Near to Dictatorship," *The Guardian* (UK) March 13, 2006.

12. George Orwell, *Nineteen Eighty-Four* (Penguin Books, Harmondsworth, UK, 1954).

13. Donald W. Livingston, "Decentralists or DC Centralists? Overthrowing the Tyranny of Liberalism," *Chronicles*, April 1999, pp. 16–18.

14. Scott Manning, "Communist Body Count," December 4, 2006, at <http://www.digitalsurvivors.com/archives/communistbodycount.php>.

15. Harold Pinter, Nobel Lecture, "Art Truth and Politics," 2005, at <http://nobelprize.org/literature/laureates/2005/pinter-lecture-e.html>.

16. John Kekes, *Against Liberalism* (Cornell University Press, Ithaca, 1977), p. ix.

17. Radicals have also seen liberalism as something of a social acid. Mao Tse-tung said, "liberalism rejects ideological struggle and stands for unprincipled peace, thus giving rise to a decadent, philistine attitude and bringing about political degeneration... It is a corrosive which eats away unity, undermines cohesion, causes apathy and creates dissention." Mao Tse-tung, "Combat Liberalism," in A.K. Bierman and J.A. Gould (eds.), *Philosophy For a New Generation* (Macmillan, New York, 1973), pp. 449–450.

18. *Gayssot Act* (France), Article 1, 1990, at <http://www.legifrance.gouv.fr/WAspad/Untex teDeJorf?numo=JUSX9010223L>.

19. Ibid.

20. H.O.J. Brown, "Cultural Revolutions," *Chronicles,* June 2001, pp. 6.

21. J. Philippe Rushton, *Race, Evolution and Behaviour,* 3rd edition (Charles Darwin Research Institute, Port Huron, 2000).

22. Wayne Lutton and John Tanton, *The Immigration Invasion* (The Social Contract Press, Petoskey, 1994).

23. Brian Appleyard, *Understanding the Present: Science and the Soul of Modern Man* (Picador/ Pan Books, London, 1992), p. 236.

24. George Orwell, *Animal Farm* (Penguin Books, New York, 1951), p. 114.

25. On the contradictions between the liberal doctrines of feminism and multiculturalism see S.M. Olin, "Feminism and Multiculturalism: Some Tensions," *Ethics,* vol. 108, 1988, pp. 661–684, and J. Cohen, M. Howard, and M.C. Nussbaum (eds.), *Is Multiculturalism Bad for Women?* (Princeton University Press, Princeton, NJ, 1999).

26. J.S. Mill, *On Liberty* (Longmans, London, 1874).

27. James Burnham, *Suicide of the West,* p. 139, from note 6.

28. Stephen Boyden, *The Biology of Civilisation: Understanding Human Culture as a Force in Nature* (University of New South Wales Press, Sydney, 2004).

29. J.W. Smith, et al., *The Bankruptcy of Economics* (Macmillan, London, 1999).

30. Anthony Barnett, "Bush Attacks Environment 'Scare Stories,' " *The Observer,* April 4, 2004, at <http://observer.guardian.co.uk/international/story/0,6903,1185292,00.html>.

31. Ibid.

32. P.K. Feyerabend, *Against Method* (SCM Press, London, 1975); P.K. Feyerabend, *Killing Time* (University of Chicago Press, Chicago, 1995).

33. A. MacIntyre, *After Virtue* (Duckworth, London, 1981).

34. See D. Greschner, "Feminist Concerns with the New Communitarians: We Don't Need Another Hero," in L. Green and A. Hutchison (eds.), *Law and Community: The End of Individualism* (Carswell, Toronto, 1998), pp. 124–125.

35. Lawrence Dennis, *The Coming American Fascism* (Harper and Brothers Publishers, New York, 1936).

36. Ibid., p. 17.

37. W. Ophuls, *Ecology and the Politics of Scarcity: Prologue to a Political Theory of the Steady State* (W.H. Freeman, San Francisco, 1977).

38. W. Ophuls and A.S. Boyan, Jr., *Ecology of the Politics of Scarcity Revisited: The Unravelling of the American Dream* (W.H. Freeman, New York, 1992).

39. W. Ophuls, *Requiem for Modern Politics: The Tragedy of the Enlightenment and the Challenge of the New Millennium* (Westview Press, Boulder, CO, 1997).

40. Ibid., p. 1.

41. Ibid., p. 45.

42. Ibid., p. 56.

43. Ibid., p. 56. See also J.R. Saul, *The Collapse of Globalism and the Reinvention of the World* (Penguin Group, Melbourne, 2005).

44. John O. Edwards, "Gen. Frank Doubts Constitution Will Survive WMD Attack," NewsMax.com, November 21, 2003, at <http://www.newsmax.com/archives/articles/2003/11/20/185048.shtml>.

45. Ibid.

46. John O'Connor, "Are Civilisations Clocks Running Backwards?" *The Independent Australian*, winter 2004, pp. 22–23.

47. C. Johnson, *The Sorrows of Empire* (Verso, London, 2004).

48. Ibid., p. 310.

49. For discussions see S. Brittan, "The Economic Contradictions of Democracy," *British Journal of Political Science*, vol. 5, 1975, pp. 129–150; Robert Kaplan, "Was Democracy Just a Moment?" *The Atlantic Monthly*, December 1997, pp. 55–80; P. Kelly, "Can Democracy Survive?" *The Australian*, May 30–31, 1998, pp. 25–26.

— 8 —

Is There an Authoritarian Alternative?

Dictatorship naturally arises out of democracy and the most aggravated form of tyranny and slavery out of the most extreme liberty.

—*Plato*

THE CRISIS OF CIVILIZATION

Environmentalist writers have had a love affair with democracy. Numerous texts have outlined the perils that the planet faces, only to conclude in the final chapter that all will be well with more democracy and a world parliament,[1] or with the creation of direct democratic communities, locally self-sufficient and living in harmony with their environment. These warm, cozy, and politically correct worlds would no doubt be a joy to live in, but they are far from the likely realities that we face. Before we can outline what sort of system we ought to have, we need to know what the likely end result will be of the dangers described in this book.

The most pessimistic response to the "crisis of civilization," that is, the multitude of interconnected social, technological, and environmental problems that humanity faces, is human extinction. The Canadian philosopher John Leslie in his book *The End of the World* takes that view.[2] Leslie considers humanity to be more at threat from technological disasters such as nuclear war, the rise of intelligent robots, and asteroid collision, than from mundane threats such as water shortages, soil erosion, and climate change.

His view is very much a technical logician's view of reality. It would take us into too many technical matters to rebut Leslie's view firsthand here. Generally, his critics seem to have established that, apart from four science fiction scenarios (killer robots, runaway high-tech experiments with exotic matter, the creation on earth of black holes, nanotechnology "grey goo" problems, etc.), none of the scenarios sketched in his book will exterminate all human life. However these scenarios will destroy the present world as we know it and necessarily cull the present human population of over six billion.[3]

Consider but one of the problems that we have discussed: the end of cheap oil. Suppose that the school of thought of the oil limitationists is right. Some estimates of the date of peak oil production put this at the year 2008, others at 2012, still others somewhat later, but many experts believe that this date will be before the end of the second decade of this century.

Although the oil optimists hope that rising oil prices will make other fuels competitive and that by market forces other substitutes will replace oil, this process will only occur if there really are substitutes. There are limits to all other forms of energy, such as nuclear fission and solar energy.[4] Even if there was an oil substitute, there would need to be a replacement of the oil infrastructure—and our civilization could not exist without oil. Plastics are made from it, and there could be no computer-based society without plastics. The world's 500 million cars depend upon oil; so does agricultural food production through fertilizers and pesticides. Coal and natural gas offer only a stopgap measure, as these reserves will also deplete—at the price of perhaps making the earth uninhabitable through global warming. Coal is mined using machinery that uses oil, and the extraction of coal will become increasingly expensive.[5]

Without a replacement of the oil infrastructure, social chaos is likely. For example, the globally connected information economy depends upon an abundant and secure supply of electricity. Without it, the security of the power grid is threatened, and with it goes the information economy. Indeed, even regular blackouts could have major economic impacts, as the August 2003 outage in the United States showed. Likewise our agricultural systems face collapse from the same dilemma. The problem of depletion is made much worse of course by the vested interest in the oil society not to seek alternatives with the same level of anxiety that one would approach a war. Even from an optimistic viewpoint, oil reserves will decline and the price of oil will soar. There is no comprehensive alternative in sight, so that even if civilization will not collapse, at least this is a matter of the gravest concern. As we have seen, there is an inertia in liberal democracies that prevents governments dealing with long-term threats. Any government that acted to curb even one use of oil by the voting citizens of a liberal democracy would be thrown out of office. If we are realistic and honest we must conclude that the inertia of liberal democracies will ensure that the problem of oil depletion is not solved before it is too late.[6]

Yet already the oil depletion problem has produced, at least in part, two wars in the Middle East and restrictions of civil liberties through laws such as the U.S.A. Patriot Act. The U.S. desire for oil reserves led the United States to support Saddam Hussein in the Iran/Iraq war and Osama bin Laden in the Afghanistan war against the Soviet Union. The United States then waged two wars in Afghanistan and in Iraq.[7] The United States presently sends a quarter of all its exported military weapons to Saudi Arabia, a regime that is at least as oppressive as Iraq was and probably will remain so. Some have argued that the United States supports Israel in the Middle East because of the push of an extremely powerful Jewish lobby in the United States and also historically because Israel served as a bulwark against what was thought to be a Sovietization of states such as Egypt, Iraq, Syria, and Yemen. Former Malaysian Prime Minister Mahathir Mohamad in an open letter to the American Muslim community has said: "In Palestine, Israeli gunship and tanks razed villages and towns to the ground, killing innocent men, women and children."[8] Some have argued that the U.S. support of Israeli human rights violations is one of the key issues that have made the United States a target for Islamic terrorists. Israelis argue in reply that Palestinians violate Israeli human rights through suicide bombings and terrorism and that Israelis have a right to self-defence.[9]

According to former Secretary of Defense Donald Rumsfeld, because of the 9/11 attacks the United States had embarked on a "thirty to forty year war against fundamentalist Islam."[10] The CIA predicts that terrorists are likely to explode a nuclear bomb on a major U.S. city such as New York in the next 20 years. New York is thought to be the target because of its high Jewish population. Osama bin Laden in his first tape released after 9/11 stated that one of the reasons for the attacks was to punish the United States for its support of what he saw to be Israel's oppression of the Palestine people, while others see this as mere rhetoric.[11]

As we stated in our last chapter a major terrorist attack on a U.S. city using a weapon of mass destruction would likely lead to martial law. Already under the U.S.A. Patriot Act a person can be arrested without probable cause and detained indefinitely without being charged. Imagine then the measures that would be put in place to save the system when the power elites are *really* under threat.

Therefore it is reasonable to suppose that liberal democratic structures will be abandoned by the existing states in an attempt to deal with the crisis of their civilization. More authoritarian structures than exist at present will arise. This, we contend, is the most reasonable inference to make from the facts discussed in this book. We predict that these authoritarian structures will be put into place to preserve the decaying status quo, rather than to begin to forge a new system of governance. It would constitute a radical historical discontinuity if this was not so, for throughout human history when those in power are under threat, they have always held on until the bitter end. Then, they are usually replaced by force.

LEAVE STALIN AND HITLER DEAD

In proposing that liberal democracies will be replaced by authoritarian structures, we differ somewhat from a select group of environmentalist writers who have also rejected a liberal democratic solution to the environmental crisis.[12] In general, such writers have felt that only centrally commanded economies can meet the challenge of dealing with the environmental crisis. We do not join that camp. We recognize that command economies committed to militarism and industrialization can be just as destructive, if not more so than liberal democracies. The former Soviet Union is not our idea of paradise on earth. Planned economies, where there is an attempt by a body of elite planners to coordinate all aspects of an economy, is a recipe for disaster because there is simply too much information, chaotic nonlinear effects, and unpredictable events to permit accurate planning. However we believe that many aspects of the economy must be firmly regulated. This position is a long way away from a planned economy.

We have no lingering belief that communism could or will save humanity, but we hold that when civilization-threatening changes occur, liberal democratic solutions are the first things to go. The rule of law is abandoned, and the rule of the strong dominates. We are not indicating that we like this; we are maintaining as a matter of real politick that this is what occurs historically and is likely to occur again. Nor are we supporting a form of authoritarianism as witnessed in Nazi Germany where one Fuhrer makes fundamental decisions about life and death for society. Such forms of authoritarianism typically lead to social disaster when the leader, following the weaknesses of human will, succumbs to corruption or madness. Our form of authoritarianism looks to the leadership of an entire stratum of society rather than one individual or even party. and there is a better chance that corruption and madness of the Hitler and Stalin levels can we weeded out. But there is no guarantee; human life is uncertain and down the track, human life promises to be desperate.

Thus unlike other antidemocratic theorists from Plato on, we do not have an alternative political ideology that we wish to promote in the place of liberal democracy, beyond that of environmentalism. We have no vision of a set of wise liberal leaders sitting in the wings, waiting to ride onto the set just in the nick of time to save us all by democratic means. Rather we have a stark vision of liberal democracy being destroyed by its own internal contradictions, in the process being replaced by authoritarian structures. It is important therefore to ask whether there exist any state authoritarian structures that are worthy of consideration. We believe that Singapore falls into the category.

SINGAPORE AND "ILLIBERAL DEMOCRACY"

In the face of the environmental failures of the liberal democracies, there may be lessons to be learned from one country, Singapore, which is often called authoritarian and an "illiberal democracy." Singapore became independent in 1965 when, like many other countries in the third world, it was poor and lacked natural resources.[13] Today its citizens have one of the highest per capita incomes in the world without suffering the sectional and social consequences of affluence. Yet Singapore is in effect a one–party state with minimal parliamentary opposition and restrictive laws. The People's Action Party (PAP) was elected in 1959 and has governed ever since. It has dedicated itself to economic success by value creation and full employment. It has created high standards in management, housing, health, education, transportation, and the environment. It has used the expertise of multinational corporations without succumbing to their philosophy. When the PAP is in effect the state, why has its authoritarian rule not become corrupt and incompetent?

Lee Kwan Yew was the leader of the PAP in the first decades of its rule. He was a highly intelligent technocrat who avoided the cult of personality and established a team based on intellectual and technical ability. Government is a meritocracy that renews itself from within its own ranks. Transition of leadership is managed carefully and appropriately without the vituperation and denigration so prevalent in the liberal democracies. Economic advancement has been a legitimizing factor for authoritarianism and opposition is insignificant. In the sphere of parliamentary opposition there are nominated members to represent particular interests and expertise.

The PAP did not evolve into an authoritarian structure. It was created in this mold. Lee said that the PAP founders "believed that political stability was the top priority because it was a prerequisite for development and modernization. This belief accompanied a shared apprehension about the transferability of Western democracy to an Asian society and an underlying conviction that unfettered democracy contained within it certain frailties always threatening to degenerate into mob rule."[14] This viewpoint from an Asian culture reflected Plato's conclusions from centuries before and has been justified by Lee's outcomes.

Singapore demonstrates that it is possible for a state to fashion an intellectual elite that can succeed in creating a wealthy economy for all its citizens. In doing this it does not allow the freedoms that many self-proclaimed leaders of the world's liberal democracies enjoy. However the freedoms of democracy are increasingly eroded by leaders using the threat of terror and the imposition of law and order to bolster their own power. It is becoming debatable whether it is better to live under these deprivations or under a benign authoritarianism that provides basic human needs necessary for well-being. Let us take the argument further by asking whether a Singapore system could be developed to drive vital environmental outcomes in the interests of humanity's future? The answer is surely yes. Governance is by a team of technocratic elites supported by educational structures described in the next chapter. An analysis of the pathetic, self-serving performance of many elected representatives of liberal democracies is a cogent argument for this option.

Unfortunately there is no possible progression of a Singapore system into an international sphere when the UN generally reflects only the quarrelsome, partisan interests of its nation states. So at present there is no hope of concerted action to address the crisis. However there is a glimmer of hope in the attitude of more localized governance systems. In the United States many states and cities have recognized the indolence of the Bush government and have begun to implement significant greenhouse reduction measures. Similar actions are being instituted by Australian states and by members of the European Union. These developments gel with the thesis that future society, of necessity, will be the antithesis of the globalized world. There will be local production, consumption and power generation, and environmental measures will be encompassed in these local systems of governance, which may accept the need to proceed with the effective use of authoritarian means to ensure the viability of their communities.

RELIGION TO THE RESCUE?

If, as seems likely, liberal democracies crumble under the stress and disorder of the environmental crisis, military rule is a likely outcome. But this is unlikely to be stable in the sense of preserving the consumer and growth economies. Thus these economies are likely to collapse with social order being held by military force for a period of economic decline that in turn threatens even military rule. So military force is only an effective social glue if there is ample money to pay soldiers. We conjecture that in the social chaos of the future, even this will come under threat. Societies only hold together if there is social glue, as Perelman has argued.[15] When the social glue of the religion of consumerism dissolves, what could replace it? Humanity has traditionally only one answer: religion. Perelman speculates that by the late twenty-first century, liberal democracy will be replaced by a form of feudalism with a steady state economy centered on land, social stratification by caste or class, and a theocracy. Removal of cultural materialism will create a void in people's lives (i.e., those surviving) that has been traditionally filled by religion. In modern times this void has been filled by materialism and science in the West, but in places where materialism and science are lacking, such as much of Africa, traditional religious practices such as voodoo continue. Today, in the West, the traditional Christian churches have declined, though some evangelical sects have increased their numbers at the expense of traditional churches. The apparent rise in Christian fundamentalism in the United States is due to the mobilization and increasing vocality of the existing numbers.[16] Fundamentalist Islam is the fastest growing religion in countries such as the United States, England, and France. These religions offer a comprehensive alternative to materialism, and when crises hit and the present social order is shattered, it is natural that people will seek refuge in "other worldly" religions.

Fundamentalist Christianity and Islam fit better with an authoritarian social structure and a steady state economy than with liberal democratic institutions. However they are not the only contenders for providing social glue for the masses. Although too much of the natural world will be destroyed for civilization to continue in its present form, some biodiversity will still exist. That which is scarce and life-giving is a fit object for valuing. Then, to live in the modern concrete jungle will make the savannah and woodlands, our instinctual home, seem more precious. It is not impossible that from the green movement and aspects of the new age movement a religious alternative to Christianity and Islam will emerge. And it is not too difficult to imagine what shape this religion could take. One would require a transcendent God who can punish and reward—because humans seem to need a carrot and stick. This God though has no chosen people. IT created the universe and earth to be diverse and rich with life: The point of life being to preserve biodiversity and allow evolutionary processes to celebrate continuous creation by producing increasingly complex lifeforms. This sacred process has been abused by humans. Those who continue to harm nature will be recreated to live in the obsolete concrete jungle, the hell of this reformed religion. The believers will live in peace and ecological fraternity in a new Garden of Eden—but perhaps this is what Christianity was all about anyway? All of this discourse will need to be written in fine prose and poetry and delivered to a prophet (rather than a profit) by "God."

Absurd, the scientific rationalists would say. Of course it is absurd, nonsense, foolishness! But so surely is the modern interpretation of Christianity, Judaism, and Islam! Yet these religions have survived while the social orders in which they existed have collapsed or changed. If the predictions of this book are correct, our social order too will be heading for the scrap heap of history. Religion will be retrieved from its slow decline in the secular Western nations. Take your pick about which religion you would like to live under, especially if you are female. Surely it is better, if we are to live under an authoritarian regime in a theocracy, that we begin to define the type of theocracy under which we wish to live. If there are no live contenders to the Abrahamic religions of Christianity, Judaism, and Islam, then a repeat of the past is what we will get.

It is not our aim to explore the question of the future of religion in any depth. An inverse relationship between religiosity and materialism has been postulated, the rich countries being secular and the poor countries steeped in religion, but clearly this does not apply to the United States. And within the rich countries, we note that the predictions of humanists and atheists that religion will fade away with increased technological sophistication have not been met. It is predictable therefore that with the collapse of a materialist culture, religion will become even more relevant in people's lives.

We have followed Perelman in maintaining that after the collapse, liberal democracy will be replaced by a form of feudalism. There will be a steady state economy based upon land and natural resource use, social stratification by caste or class or perhaps ethnicity, and theocratic rule. We turn now to develop the thesis in more detail and reply to some anticipated objections.

A NEW FEUDALISM?

The debate that we hope to initiate is quite removed from the one focused on Fareed Zakaria's *The Future of Freedom: Illiberal Democracy at Home and Abroad*.[17] Zakaria argues that developing societies fare better (i.e., economically) under liberal authoritarian regimes. The question is part of a debate about the best approach for national development in the developing world for the attainment of economic growth. Obviously enough, for those who have read this far, the assumption of continuous economic growth is challenged by a vast amount of environmental evidence. Nevertheless Zakaria makes some important observations about how countries like the United States, which have overdosed on democracy, have ultimately eroded basic liberties. He also wisely observes that democracy in the Arab world is likely to produce rulers more like Osama bin Laden than King Abdullah of Jordan. This thesis has also been convincingly argued for by Amy Chua in *World on Fire: How Exporting Free Market Democracy Breeds Ethnic Hatred and Global Instability.*[18] In the previous chapter we argued that liberalism, not merely democracy, if pushed to its logical conclusion, had illiberal ramifications.

Feudalism is wrongly associated with slavery and despotism. Despotism and oppressive, exploitative rule are neither historical nor logical consequences of an authoritarian rule. Slavery flourished in both democratic Athens and the American South, prior to the Civil War. Feudalism in Europe was an authoritarian system but it lasted longer than capitalism is likely to last. As Daly and Cobb have observed,[19] it was more a communitarian system than either capitalism or socialism. John Stuart Mill in his *Principles of Political Economy* noted that feudalism of the late Middle Ages produced in England a "yeomanry who were vaunted as the glory of England while they existed, and have been much mourned since they disappeared."[20] Even the neoclassical economist Alfred Marshall was moved by a consideration of feudal society to note that "in the Middle Ages…the great body of the inhabitants frequently had the full rights of citizens, deciding for themselves the foreign and domestic policy of the city, and at the same time working with their hands and taking pride in their work, they organized themselves into Guilds, thus increasing their cohesion and educating themselves in self-government."[21]

A defense of the guild system and its use after the collapse of capitalism is given by Arthur J. Penty in *Guilds and the Social Crisis.*[22] A general debunking of the myth of the dark oppressive nature of medievalism is given by Hilaire Belloc in *The Servile State,*[23] L. Kohr, *The Breakdown of Nations,*[24] and Peter Laslet, *The World We Have Lost.*[25] Most importantly for those who are concerned about the environmental crisis, the mathematical economist Nicholas Georgescu-Roegen[26] has shown that feudalism allows more people to exist than under capitalism in conditions of overpopulation and low productivity—the likely state of future humanity. We point out that each generation has the psychology to want to believe that it has improved the lot of humanity and in particular, today, with the comforts of materialism, life must be the best it has ever been. But we are writing as the articulate, comfortable, employed, and feudal England might have attractions for today's poor, unemployed, and homeless.

However, in our opinion it is highly doubtful that the same bonds and reciprocation of loyalty as characterized European feudalism would revive in the new social order. It is difficult to make any reasonable predictions about this social order beyond the generalities that it will have a steady state economy centered on land, that it will be stratified rather than egalitarian, and that there will need to be some type of social glue, a role traditionally served by religion before its replacement by materialism.

The hypothesis that a steady state economy will characterize the new order follows from the limits to growth thesis defended in this book. If a growth economy is not sustainable, then the economy must be either a steady state nongrowth economy or one that is constantly decreasing and degenerating. A degenerating economy eventually leads to economic collapse that is nonsustainable. By elimination, a sustainable economy must be a steady state economy.

A future society is likely to be stratified and nonegalitarian because history shows that this is the way societies in the past have been. The hypothesis defended in this book is that liberalism and its values, as well as democracy, are just moments in human history. It is likely that the human brain is hard-wired for authoritarianism, for dominance, and submission (chapter 5). This is a reasonable scientific hypothesis that better fits the available historical evidence than the hypothesis of liberal egalitarianism.

Finally, let us consider religion as alternative social glue for society. Firstly, sociology is premised on the idea that social structures require a glue or bonding material to hold them together. The building metaphors are appropriate. No human society has yet been found without cultural beliefs that tie individuals together. This idea seems to be conceptually bonded to the very idea of social order and certainly all of the founding fathers of sociology—Karl Marx (1818–1883), Emile Durkheim (1858–1917), and Max Weber (1864–1920)—thought so. Secondly, some have speculated that humans also have a genetic predisposition toward religious faith. Dene Hamer in *The God Gene*[27] identifies one gene as responsible. Although it is unlikely that the picture of genetics of spirituality is as simple as that, there is evidence to suggest that there is a genetic basis for spirituality. Twins separated at birth, despite different upbringings and environments, tend to have similar levels of spirituality and identical twins (monozygotic twins) are twice as likely to have a similar level of spirituality as fraternal or dizygotic twins.[28] Religion has provided social glue for human society in the past and it is reasonable to hypothesize that it will do so again with the future necessary demise of the consumer materialistic lifestyle.

Therefore all of the basic components of a feudal society are predictable on the basis of the arguments of this book. We do not discount totally the possibility of some high-tech technocracy arising where thinking machines could rule the world. However we believe that all of the science fiction scenarios in popular movies (e.g., *Terminator III, I Robot,* etc.) presuppose technological advances that are not likely to be made before the end of industrial society. Humankind is therefore unlikely in the short term to be able to download its consciousness into android form and escape a biological and ecological fate.

The ultimate shape of future human societies depends upon the time frame adopted by humanity for dealing with the environmental crisis. On a business as usual scenario it is most likely that a great dying will lead to life as seen in the failed African states of today with a rule by warlords. There are an infinite number of less pessimistic scenarios.

In this situation it is wise not to make any definite predictions about the shape of future human society beyond the generalities that have been made already. However in the remainder of this chapter and in the next we will broadly outline what an authoritarian government ideally *should* be like.

MEET THE ELITES

It is foolish to attempt to sketch any detailed model of what an ecologically sustainable authoritarian government would be like. We cannot anticipate the full scope of the environmental damage to which the planet will be subjected before humans wake up, if they do so at all. However some broad generalizations can be made at this point in history.

We propose that any sustainable society, even if it takes the form of a group of tribes living in a state depicted by the *Mad Max/Road Warrior* movies will be centered around ecology rather than economics. Concerns will be biologically based rather than consumption based. This will become the necessity. Recognizing the move to conflict and war for environmental resources (chapter 3) we emphasize the need for structures that utilize peaceful mechanisms. However attractive to us as near primates, the guerrilla methods of *The Monkey Wrench Gang*[29] must be denounced.

We advocate a form of governance by authoritarianism abhorrent to liberal thinkers. But society is already managed by the hidden hand of the financial elite, and freedom is illusory and diminishing. Little can be done about the fact that we, the ordinary people, will wear chains, as we have always done. But perhaps the type of chains and how tightly they bind us can be influenced by our thinking.

We commence with a description of the elites that we don't want; it is then possible to see the flip side of this character. Government today is primarily influenced by economic policies and modes of thought and is executed by the elected politician who with very few exceptions has emerged as adept at working corrupt party machines. Those who out-maneuver their colleagues to gain leadership are reluctant to leave and often begin the inevitable moves to authoritarianism. They are universally poorly regarded by the electors, and it is worthwhile to quote the insightful remarks of that brilliant iconoclastic writer and political commentator H.L. Mencken (1880–1956) who observed that politicians

seldom if ever get there [into power] by merit alone, at least in democratic states. Sometimes, to be sure, it happens, but only by a kind of miracle. They are chosen normally for quite different reasons, the chief of which is simply their power to impress and enchant the intellectually underprivileged... Will any of them venture to tell the plain truth, the whole truth and nothing but the truth about the situation of the country, foreign or domestic? Will any of them refrain from promises that he knows he can't fulfill—that no human being *could* fulfill? Will any of them utter a word, however obvious, that will alarm and alienate any of the huge pack of morons who cluster at the public trough, wallowing in the pap that grows thinner and thinner, hoping against hope? Answer: maybe for a few weeks at the start... But not after the issue is fairly joined, and the struggle is on in earnest... they will divest themselves from their character as sensible, candid and truthful men, and become simply candidates for office, bent only on collaring votes.[30]

Hans-Hermann Hoppe in his book *Democracy: The God that Failed* develops Mencken's critique of democracy in this respect.[31] He says that democratic popular elections make it impossible (difficult, we believe) for good and decent people to rise to the top. Much like a pot of boiling water containing impurities, the scum will rise to the top. And as we have seen, they do so. Hoppe, with his typical tough turn of phrase, laments that under democracy the leaders are increasingly bad and sadly only "rarely assassinated."[32]

The democratic system itself attracts to politics those people who are the most unsuitable for government. We should add that most authoritarian systems are also defective in this respect. The ruling elites typically first obtain power by violence, usually in the midst of the breakdown of democracy. An oppressive state machine is then set up and perpetuated by megalomaniac types who lust for power as a mode of personal advancement.

Have we set ourselves an insurmountable problem? Who are to be the new elites? In capitalist society, where money and self-promotion rule, they are invisible. Since we wish to avoid self-selection, how are they to be drafted into service? Intellectualism alone is not sufficient, for in the past century the intellectual has succumbed to the hymns of tyranny as often as the rest of us.[33]

Perhaps we could commence by identifying those leaders in history who were humble and worked for the common good. Yes, there were some who did not fit the selfish mores of society. We crucified them, ignored them, burned them, or, in modern times, shot them. Jesus Christ, Buddha, Socrates, St. Francis of Assisi, Dali Lama, and Gandhi. The difficulty of the task indicates the inadequacies of humankind.

Let us take the question a stage further. Are there any individuals, not interested in self-aggrandizement and accumulation of material assets, who have broad intellectual, scientific, and social managerial skills to lead humanity through the environmental crisis? By definition, they have not placed their head above the parapet to join the scramble of economic rationalism. Such persons of integrity and learning have been sought for centuries. Aristotle referred to them as aristocracy. This meant "the best," as interpreted by Graham in *The Case Against the Democratic State*.[34] They were those with the abilities and attitudes of mind to be entrusted with government. The sixteenth century philosopher Etienne de la Boetie said in his treatise, *The Politics of Obedience,* the following:

> There are always a few, better endowed than others... These are in fact the men who, possessed of clear minds and farsighted spirit, are not satisfied, like the brutish mass, to see what is at their feet, but rather look about them, behind and before, and even recall the things of the past in order to judge those of the future, and compare both with their present condition. These are the ones who, having good minds of their own, have further trained them by study and learning. Even if liberty had finally perished from the earth, such men would invent it. For them slavery has no satisfaction, no matter how well disguised.[35]

Both de la Boetie and Hoppe are primarily concerned with the preservation of freedom of the individual, this being the core value in their systems. But for us freedom is not the most fundamental value and is merely one value among others. Survival strikes us as a much more basic value. Now our proposal is that since fighters for freedom are always likely to arise, the probability of fighters for life and survival arising must be as great if not greater. This will be especially so if the opportunity is provided for such ecowarrior/philosophers to develop and be nurtured in special institutions called "real universities" or academies. At present our leaders are primarily trained in institutions that perpetuate and legitimate our environmentally destructive system. The conventional university trains narrow, politically correct thinkers who ultimately become the economic warriors of the system. Our proposal is to counter this by an alternative framework for the training and complete education of a new type of person who will be wise and fit to serve and to rule. Unlike the narrowly focused economic rationalist universities of today, the real university will train holistic thinkers in all of the arts and sciences necessary for tough decision making that the environmental crisis confronts us with. These thinkers will be the true public intellectuals with knowledge well grounded in ecology. Chapter 9 will describe in more detail how we might begin the process of constructing such real universities to train the ecowarriors to do battle against the enemies of life. We must accomplish this education with the dedication that Sparta used to train its warriors. As in Sparta, these natural elites will be especially trained from childhood to meet the challenging problems of our times.

Government in the future will be based upon (or incorporate, depending on the level of breakdown of civilization) a supreme office of the biosphere. The office will comprise specially trained philosopher/ecologists. These guardians will either rule themselves or advise an authoritarian government of policies based upon their ecological training and philosophical sensitivities. These guardians will be specially trained for this task.

In the meantime can we move forward? There *are* those unsullied by the search for power and influence and with ability, who already serve humanity with meager financial reward in the professions, in science and medicine, in the entrepreneurial social services, and, yes, in the religious orders, but they are not prepared to join the political rabble. And even if they were, the present political cabal would not move over for them. The emergence of the World Social Forum may offer some lessons in networking individuals with common goals, the seed of an international organization for environmental equity and sustainability. This would not be inward looking and self-serving like various Zionist organizations or the Yale Skull and Bones, but might be universal like a reformed Roman Catholic Church. Come back St. Francis!

Authoritarian leadership exists in the Roman Catholic Church where power and greed are successfully suppressed to deliver spiritual succor to the believers and nourishment for the poor. Lessons can be learned from the modus operandi of this Church. In its service to humanity it publicly abhors the destructiveness of both totalitarianism and capitalism, and its views might allow it to be the chrysalis of care for the earth through directions to its flock. As Pope John Paul II stated:

The ecological crisis is a moral issue... respect for life and for the dignity of the human person extends also to the rest of creation... Humanity has disappointed God's expectations. Man, especially in our time, has without hesitation devastated wooded plains and valleys, polluted waters, disfigured the earth's habitat, made the air unbreathable, disturbed the hydrological and atmospheric systems, turned luxuriant areas into deserts and undertaken unrestrained industrialization... We must therefore encourage and support the ecological conversion which in recent years has made humanity more sensitive to the catastrophe to which it has been heading.[36]

A recurrent theme in this text is the need for a new religious basis to modern life to give substance and meaning to people's existence as an alternative to consumerism and materialism. A "green pope" who actively pursued the philosophical words of Pope John Paul II quoted above would make a substantial contribution to the saving of civilization. But there is another important contribution that Catholicism offers to our argument. The Roman Catholic Church is one of the longest surviving Western institutions. It is much older than the common law, democracy, the English language, and Western science. The Church has seen the collapse of one civilization (the Roman), has existed through a dark age, and survived wars, revolutions, and plagues. As a social institution it is truly remarkable and offers to all of us a lesson in how to set up an organization for long-term survival.

What is important for our argument is that the Roman Catholic Church, unlike the fragmented Protestant churches, has a rigid authoritarian structure and a strict hierarchy of rule. If the Roman Catholic Church had been run as a democratic institution, as the Protestant churches have been to some degree, it is highly doubtful whether the Roman Catholic Church would have survived. We do not see in the Church an exact model to replicate for an alternative authoritarian model of government, as it obviously would be a dangerous gamble to have one person as a "political pope" or world emperor. Nevertheless the survival of the Church as an authoritarian structure does indicate that authoritarian systems, if set up correctly, can be long lasting and stable.

Our discussion would be incomplete without recognizing the most successful authoritarian systems in human history, the corporations. These systems have immense financial power and influence and increasingly hold the nations in a grip from which they cannot escape. They toy with democracies as cats with mice and then dispatch them when necessary. We have argued that, on balance, their existence is not in the service of humanity, though with their delivery of comforts they may delude themselves that it is. In the sphere of our discussion of the need to sustain the natural environment, they continue to be a force for evil. These authoritarian structures are strictly pyramidal and direct their obedient workers to a unified goal, the pursuit of profit. They are immensely successful in what they do, so much so that the assets of one may exceed those of many countries. As a result they are at the top of the food chain and Plato's mob will never control them. All else is subservient. Perhaps only an authoritarian system with unified goals for the future of humanity can control them for the common good.

What then will be the structure of the authoritarian government by the elites? In chapter 1, we alluded to authoritarian management structures like those of the hospital's intensive care unit where personal power, financial gain, and personal aggrandizement are subsumed into the goal of saving human lives. Here is a multidisciplinary team of experts working for humanity without fear or favor. It is a model for the future of nations and the world.

It is not possible to take the argument further. Today we are reluctant to add the names of any individuals who could be conscripted for our alternative "Intensive Care Management Government," because there are obviously defects in all individuals educated in our existing institutions—including us! Nevertheless, as Darwinian evolutionists we believe imperfections can be eliminated by a process of trial and error and selection. We can rebuild the ship of civilization while it floats, slowly attempting to produce better qualified people, people who are less selfish and more altruistic than ourselves.

The time frame for any sort of education-based leadership change will be many decades and of course, humanity does not have the luxury of waiting for such a time. Therefore, in our opinion, there is a considerable likelihood that some type of economic or ecological crash will occur that will lead to the collapse of our present social system. There will thus be casualties; there is no escape from the fact that a great reckoning for humankind is to come. What we propose is a form of crisis care management so that civilization does not perish; we wish to save a remnant.

Of course we have not answered all the questions that naturally arise when any strategy of "how to get there" is postulated. Given that there is so little thought about what to do in such worse case scenarios, we believe that some process is better than nothing at all. Given the problems we have sketched, it is difficult to see where else one could go or what else one could do. Therefore, take our proposal as a "work in progress" research program that can be developed further.

NOTES

1. George Monbiot, *The Age of Consent. A Manifesto for a New World Order* (Flamingo, London, 2003).

2. John Leslie, *The End of the World; The Science and Ethics of Human Extinction* (Routledge, London, 1996).

3. See J. W. Smith, et al., *Global Anarchy in the Third Millennium* (Macmillan, London, 2000).

4. David Goodstein, *Out of Gas: The End of the Age of Oil* (W.W. Norton, New York, 2004).

5. Ibid.

6. J. H. Kunstler, *The Long Emergency* (Atlantic Books, London, 2005).

7. M. Mendel, *How America Gets Away With Murder: Illegal Wars, Collateral Damage and Crimes Against Humanity* (Pluto Press, London, 2004).

8. Mahathir Mohamad, "Open Letter to American Muslims," October 16, 2004, at <http://www.islamicity.com/articles/Articles.asp?ref=IV0410-2488>.

9. See generally, for a strong statement against Zionism, A. Loewenstein, *My Israel Question* (Melbourne University Publishing, Melbourne, 2006), and for a strong statement of the Zionist position, A. Dershowitz, *The Case for Israel* (Wiley, New York, 2004).

10. R. Freeman, "Will the End of Oil Mean the End of America?" March 1, 2004, at <http://www.topplebush.com/oped284.shtml>.

11. Mendel, *How America Gets Away With Murder,* from note 7; Dershowitz, *The Case for Israel,* from note 9.

12. W. Ophuls, *Ecology and the Politics of Scarcity: Prologue to a Political Theory of Steady State* (W. H Freeman, San Francisco, 1977); Robert L. Heilbroner, *An Inquiry into the Human Prospect* (W.W Norton, New York, 1974).

13. D. K. Mauzy and R. S. Milne, *Singapore Politics Under the People's Action Party* (Routledge, London, 2002).

14. Ibid., p. 6.

15. L.J. Perelman, "Speculations on the Transition to Sustainable Energy," *Ethics,* vol. 90, 1980, pp. 392–416.

16. P. Norris and R. Inglehart, *Sacred and Secular: Religion and Politics Worldwide* (Cambridge University Press, Cambridge, 2004).

17. Fareed Zakaria, *The Future of Freedom: Illiberal Democracy at Home and Abroad* (W.W. Norton, New York, 2003).

18. Amy Chua, *World on Fire: How Exporting Free Market Democracy Breeds Ethnic Hatred and Global Instability* (Doubleday, New York, 2002).

19. H. F. Daly and J. B. Cobb, Jr., *For the Common Good,* 2nd edition (Beacon Press, Boston, 1989), p. 15. The authors are indebted to this text for the references in notes 20–26.

20. J. S. Mill, *Principles of Political Economy* (Kelly, Clifton, NY, 1973), p. 756.

21. A. Marshall, *Principles of Economics,* 8th edition (Macmillan, London, 1925), p. 735.

22. A. J., Penty, *Guilds and the Social Crisis* (George Allen, London, 1919).

23. H. Belloc, *The Servile State* (T.N. Foulis, London, 1912).

24. L. Kohr, *The Breakdown of Nations* (Reinhart, New York, 1957).

25. P. Laslet, *The World We Have Lost* (Scribner, New York, 1965).

26. N. Georgescu-Roegen, *Economic Theory and Agarian Economics* (Oxford Economic Papers, Oxford, 1950).

27. D. H. Hamer, *The God Gene* (Doubleday, New York, 2004).

28. Ibid.

29. Edward Abbey, *The Monkey Wrench Gang* (Harper Perennial Modern Classics, New York, 2000).

30. H. L. Mencken, *A Mencken Chrestomathy* (Vintage Books, New York, 1982), pp. 148–151, cited from H.-H. Hoppe, *Democracy: The God that Failed* (Transaction Publishers, New Brunswick, 2001), pp. 88–89.

31. Hoppe, *Democracy,* from note 30, p. 89.

32. Ibid.

33. Mark Lilla, *A Century for Tyrants,* extracted from his essay, "The Lure of Syracuse," originally published in *The New York Review of Books,* September 20, 2001; printed in *The Australian,* November 14, 2001.

34. Gordon Graham, *The Case Against the Democratic State* (Imprint Academic, Charlottesville, VA, 2002).

35. Etienne de la Boetie, *The Politics of Obedience: The Discourse of Voluntary Servitude* (Free Life Editions, New York, 1975).

36. John Paul II, "The Ecological Crisis: A Common Responsibility," World Day of Peace, January 1, 1990, at <http://www.ncrlc.com/ecological_crisis.html>.

— 9 —

Plato's Revenge

There will be no end to the troubles of states, or of humanity itself, till philosophers become kings in this world, or till those we now call kings and rulers really and truly become philosophers, and political power and philosophy thus come into the same hands.

—*Plato*

PLATO RELOADED

In the previous chapter we introduced the authoritarian position through a speculative, but we believe, rationally supportable set of predictions about the likely course of human society based upon our evolutionary understanding of human nature, as well as the recognition of the stark ecological realities that confront the human species.

We argued that the matrix of ecological, social, and political forces together constituted a potentially lethal disease for the survival of liberal society if not human civilization itself. For example, the expected decline in oil resources alone threatens the continued existence of industrial society; but this is but one problem among many in a web of problems. Climate change also threatens the continued existence of the growth economy. Consumerism and materialism, the main social glue of Western societies, depends on uninterrupted growth; but if there are approaching limits to growth then consumerism as the social glue is doomed. Liberal democratic society will quickly unravel and more quickly and more chaotically than in authoritarian regimes where populations live at more modest or subsistence levels. Western societies live at the lofty peak of the food chain, and they have a long distance to fall. We have outlined how we believe that social, political, and ecological stresses facing liberal democratic societies will work to gradually transform these societies into authoritarian regimes. We have no faith in the ultimate uprising of the masses against the forces of darkness. The 2004 elections in the United States and Australia are good illustrations that liberal democracy, like the totalitarian state, can choose to operate by promoting fear of terror or of financial damage to individuals who are primarily consumers with large debts. In neither election were the words "environmental crisis" mentioned. It is clear that the hard environmental choices cannot be made in liberal democratic societies, for the elites and the citizens of market economies have become too selfish to make sacrifices.

The question to be addressed in this penultimate chapter is what, if anything, can be done to prevent a dismal future for humanity. Stated more precisely let us add the qualifications, *utterly* and *hopelessly,* dismal future. In our view, based on all the scientific evidence available, modern society is a runaway train that must ultimately face the stark reality that at the end of the track stands the ecological mountain. There is little hope of stopping the train dead before the point of collision, but the speed of the train may be slowed to lessen the impact, and some carriages may perhaps be uncoupled from the train before the ultimate smash.

Those who reject the ideology of democracy ultimately find themselves returning to Plato's *Republic,* which was one of the most comprehensive critiques of Athenian democracy. Plato (ca. 427–347 B.C.) is regarded by many scholars as the most important philosopher in the ancient world. Like all great thinkers of that age he attempted to produce a comprehensive view of *reality,* answering questions in his dialogues through the character of his teacher Socrates, such as "what is knowledge?" and "what is justice?" For Plato statements are true or false by virtue of *reality* not by convention. Thus a true definition of justice, for example, is one which is in every respect fully just and does not vary from place to place. It is timelessly true, does not change, and is not located in any particular time or place. Plato refers to this reality of timeless truth as the "Forms." It is the philosophers who have primary access to the timeless truths of the Forms because philosophers by definition are lovers of wisdom ("philos" meaning "loving," and "sophia" meaning wisdom). Philosophers seek the truth and thus attain knowledge rather than mere belief. Philosophers love the truth and hate falsehood.

Plato compares philosophy to a ship. The master of the democracy knows nothing of navigation, so the sailors seek to control the ship. They also know nothing about navigation and sailing. The masses have no time for someone who has genuine knowledge of navigation and sailing. Cashing out the simile, Plato is proposing that government requires the sorts of skills and wisdom that philosophers possess. Knowledge of the real, rather than appearance, can only come from hard intellectual training, which only the philosophers have. Others live in the world of belief, not of knowledge, and they are motivated by self-interest rather than a love of the truth.

Plato outlined an elaborate view of his ideal society, the exact details of which need not concern us here. He had no sound mechanism for realizing this society, which he admitted had never existed, but hoped that by some chance or divine inspiration would guide philosophers to be kings. Where and when this is likely to happen is not stated by Plato.

For the modern mind almost everything of Plato's argument can be questioned. In particular, a rejection of Plato's theory of Forms as a philosophy of knowledge is sufficient to bring down the political edifice of *The Republic*. Even so, the over-rationalized view of the philosopher king hardly fits well with modern evolutionary theory and psychology. *The Republic* even for its time seems to be a fantasy. But not all elements of a fantasy need be unrealistic or unachievable.

We feel that there is some merit in the idea of a ruling elite class of philosopher kings. These are people of high intellect and moral virtue who are trained in a wide number of disciplines, ecology, the sciences, and philosophy (especially ethics), for the purpose of dealing with the crisis of civilization. Their goal will not be knowledge for its own sake, but knowledge in the service of life on earth. These new philosopher kings or ecoelites will be as committed to the value of life as the economic globalists are to the values of money and greed. In the rest of this chapter we will summarize the argument.

THE MEANING OF MEMES

Our thesis is that there is a need to establish new universities to train a new group of thinkers and activists to fearlessly tackle the problems discussed in this book. A Real University is a center of wisdom to serve the fundamental needs of humanity and nature. We have little confidence that the entrenched government/corporate enterprise will be democratically replaced, but hopefully the newly enlightened will be available when the crisis comes, a civil defense system, a government in waiting.

The concept of memes is pertinent to this discussion. The meme emerged in Richard Dawkin's *The Selfish Gene,*[1] in which it was recognized that our genetic evolution involving variation, selection, and survival of the best adapted may be accompanied by a process of copying or imitating information that is important to survival. Often meme information may give us an advantage, such as the location of the food stores or prey for primitive man, or learning how to drive defensively for modern humans. These behaviors are copied.

Memes are thought patterns in the vast information systems of our brains. They are linked to language, which may assist their formation, and they are replicated by passing them to others. They are also created by seeing interesting images and names, or by hearing catchy tunes. Such memes titillate or interest the brains of entire populations and, as with Coca-Cola, are created and disseminated by skilful advertising. Obviously it is difficult to explore memes scientifically, but they can provide a useful model for important behavioral and psychological events. For example, religious fervor has in history spread through populations via thought patterns that can dominate most other thinking, so much so that the nonbeliever may be regarded as expendable. Marxism and fascism that consumed a generation of nations and influenced every attitude of the believers are examples. Capitalism stimulates meme interest and activity in our brains through novel advertising of the consumer goods so vital for its success.

A more profound analysis of memes is self-destructive for it can lead to a conclusion that each of us is a collection of memes playing out their own interactions in our brains. There is no self, no free will, only a collection of information blown by the wind through the jungle of nerve cells in human brains and society. But let's stop before we arrive at this depressing conclusion, which would render education and life meaningless. It is fair to conclude however that fast memes have taken over from slow evolution as the means of changing society.

It is remarkable how different—and indeed alien—thinking of only a few generations ago is compared to today. At the beginning of World War I, 90 years ago, when our genetic status, compared to our long evolution, was virtually the same as today, the thoughts pervading society now seem incomprehensible. The joy, adventure, and duty of war, which embraced millions of young men and was supported by the defining principles of society, seem to defy our understanding now. Such patterns of thought resonate through society, creating a uniformity of ideas and purpose. As international information and communication systems have developed, cultural ideas, music, and images have resonated around the world in the same way.

Economic globalization and consumerism have behaved like infectious memes. They have swept through governments, bureaucracies, and corporations of Western civilization and are now infecting other cultures. The believers cannot conceive of any other system; those opposed are heretics to the cult of economics, they are retrogressive educationalists, they are political extremists of the right or left. The universities are caught in the web of this thinking. One model for universities, with only minor variations, has swept the world. It is a model for credentialism, where a degree demonstrates desirable qualities to the corporate employer such as ambition, persistence, and ability to conform and cooperate.[2] The university degree is the mechanism for winnowing and screening of applicants. The university is geared to serving industry, promoting competition and consumerism, and it squeezes out patterns of thought of no value to the cause. Its proponents are fervent, and those opposed are sidelined and replaced. Colleagues, publishers, and media quickly dismiss critical or nonconforming memes.

Yet evolution needs variation, then selection. Society has always changed and advanced in this way. Nowadays this task is much more difficult for the thinker, the academic, and the experimenter. Societies have thrived on diversity but this is now being suppressed in a society supposedly espousing individualism in thought and action.[3] To follow a need for diversity, we require alternative universities with different values and intents. There are many structural alternatives provided by online and virtual universities but these repackage the occupational needs of the consumer society. They compete for the fees of students on the world market and list themselves on the stock exchanges. We are failing the fundamental needs of humanity because we have allowed our universities to become an array of competing look-alikes aping their big brothers in the corporate world.

WISDOM: KING SOLOMON'S MIND

Each of the countless articles and books on education and its contribution to civilization should try to build upon fundamental thinking that has gone before. A Real University must examine the meaning of knowledge. This has rarely been done. But the ideas of one philosopher, Nicholas Maxwell,[4] give us a general framework for rethinking, restructuring, and reviewing the university. Maxwell links philosophy with the suffering and misery of our time. At present the acquisition of knowledge is the basis for inquiry in the humanities and social and physical sciences. What is needed, Maxwell argues, is a new kind of inquiry that gives intellectual priority to human problems and so attempts to enhance wisdom and the art of living wisely. This is a long neglected idea from the time of Socrates. When we aim to improve the quality of human life, it is profoundly irrational to give intellectual priority simply to the task of improving knowledge. Rather, we must give priority to articulating our problems of living, and to proposing and criticizing possible solutions. It is not primarily new knowledge that we need; we need to act in new and appropriate ways. As a matter of urgency we need to develop a more rigorous kind of inquiry, in many ways radically different from what we have at present, having as its basic aim to improve not knowledge but rather *wisdom*. Maxwell's views point the way forward for they prioritize human problems for investigation and action. The most urgent are peace, poverty, world health, and environmental repair.

Maxwell then distinguishes between the *philosophy of knowledge* and the *philosophy of wisdom*. The philosophy of knowledge sees the proper aim of inquiry as the acquisition of knowledge. This *may* be used to promote human welfare, but the use of knowledge is not generally of central concern to this tradition. Its concerns are intellectual rather than practical, as knowledge is the basic aim of intellectual inquiry. The fundamental description of the philosophy of such knowledge is as follows.

It is absolutely essential that inquiry must not be influenced by any kind of sociological, economic, political, moral, or ideological factors, pressures that tend to influence thought in our society. Feelings, desires, human social interests or aspirations, political objectives, values, economic forces, public opinion, religious views, ideological views, moral consideration, must not be allowed in any way to influence scientific or academic thought within the intellectual domain. Only questions of fact, truth, logic, evidence, experimental, and observational reliability and success must be considered. Only these factors can be allowed to influence the determination of truth and the acquisition of knowledge. All additional extra academic human, social consideration factors must be ruthlessly held at bay and ignored. According to this philosophy of knowledge, literature and art make no rational contribution to knowledge; they do not contribute towards truth. Essentially this viewpoint can be seen as an intellectual ego trip that has lead us to our present social crisis in civilization.

By contrast the philosophy of wisdom offers an entirely different perspective. It applies reason to the enhancement of wisdom; wisdom being understood as a desire, an endeavor, and a capacity to discover and achieve what is desirable and of value in life, for oneself and for others. Wisdom includes knowledge and understanding but goes beyond them in also including a desire and an active striving for values. Wisdom includes the ability to experience value and the capacity to use and develop knowledge, technology, and understanding for the realization of value. Wisdom, like knowledge, can be conceived of, not only in personal terms but also in institutional or social terms. We can thus interpret the philosophy of wisdom as asserting that the basic task of rational inquiry is to help us develop wiser ways of living, institutions, customs and social relations—a wiser world.

Examples of wisdom occur in every walk of life. In Western culture, the Bible relates many wise thoughts and deeds. King Solomon adjudicated on a dispute between two women over who was the true mother of a child. Each woman was adamant in claiming ownership. And the king said, "Divide the living child in two and give half to the one, and half to the other."[5] The true mother then said to give the child to the other woman to save it from death. By contrast the false mother asked for the child to be divided. By this means the king determined the correct ownership. In this judgment, King Solomon recognized and used the natural maternal response. This was not just a use of his knowledge and experience, it was wisdom and understanding of human nature, responses and emotions.

Wisdom has been translated into the environmental context by David Orr as part of "slow knowledge" that is accumulated during evolution in the process of cultural maturation.[6] It involves how to do practical things; it is the careful conservation and increase of knowledge over many generations. It constructs a society on the basis of wisdom rather than cleverness, and it recognizes that the careless application of knowledge can destroy all knowledge, for example by nuclear or biological warfare. Indeed education that is knowledge only, without wisdom, may allow people to become greater and greater destroyers of ecological services. Indigenous people display wisdom that is knowledge and the interpretation of that knowledge, passed down through generations in stories and fables. Such wisdom exists in all cultures, although we may find it more difficult to recognize it in cultures that are alien to us. But in Western cultures we recognize many instances in the professions. The practice of medicine is regarded as an art as well as a science for it requires wisdom and judgment to deliver outcomes based on medical science.

The consequences of these discussions are as follows: The freedom to pursue knowledge as the individual sees fit is a mistake, for freedom must be considered in the context of the needs of society as a whole. There must be judgment and wisdom. Freedom of research has been a holy grail for those pursuing the idea of a liberal university. It follows that the Real University will have an agenda, which includes priorities for those tasks to be pursued that are essential to the future well-being of humanity.

THE UNIVERSITY IN RUINS

These conclusions have particular relevance for science. Science is the pure pursuit of knowledge. By contrast education has a wider definition for it concerns itself not only with the pursuit of knowledge but with the promotion of social ideals and human values. Science has separated itself from its philosophical origins and needs to be reincorporated into educational practice, which includes ethics, values, and social context. Science, pursued by elite groups, has concerned itself too little with its consequences and contributions to the well-being of humanity. Its freedom has to be constrained by ethics and by observance of the precautionary principle. In addition science has to be able to provide a clear independent opinion to the public on important issues, and the community has to have means to encourage or direct science to pursue research into the major problems of the world today.

Unfortunately, the universities have failed to varying degrees on all these issues. It is too much to expect that the present university system could embrace such reforms, for it is greatly influenced by commercial reality and the present status of scientific research. The universities would need to reform the curriculum for each scientific subject to provide the student with knowledge of the origins of science, its limitations and responsibilities. The teachers would need to have a strong independent voice without fear of reprisals when they spoke on important issues. The funding system would need to provide independent funds divorced from commercial influence. There would need to be research in applied and directed areas that tackled social, community, and worldwide problems of concern to humanity. There is no evidence that any of these issues is being significantly addressed in universities today. A new structure is required in a new institution to enact these reforms and produce a balanced scientist.

For all these reasons, it is easy to agree with Bill Readings[7] that the university is in ruins, but a nostalgic return to a previous culture is impracticable and impossible. The metamorphosis and takeover of the universities cannot be unscrambled. Readings believes that we have no alternative but to embrace the corporate identity of the university. Within it, we can wander in the ruins and utilize the buildings and structures for progressive and new thinking. We can live a semiautonomous existence while paying deference to the controlling administration. The space available to us may still allow us to think and communicate. Thought is a nonproductive labor and so does not figure in the university balance sheet, except as waste. But thought must be preserved if useful activity is to be retained within the ruins.

Readings likens the threat to thought, to the progressive shrinking of the old-growth forests in the United States. This is a very good analogy. One can ask how many philosophers or redwoods are required to preserve the entity and its biodiversity. Each call for compromise between loggers and environmentalists means more forest is felled. Each interaction between administration and the academic leads to less freedom of thought in the universities. The point of no return is unclear and in environmental matters we enunciate the precautionary principle. In the universities the precautionary principle for the freedom of thought and for thought itself is not applied. In partnership with thought, those in the University in Ruins are left to participate in communication that is part of the globalized community.

Bill Readings's explanation of the transformation of the university relies on the concept that the university once had a national role that in turn created its cultural role. With globalization the national role has disappeared and the ship has become rudderless. He asks to whom or what is the university accountable? The author's philosophy is that this question no longer matters, for the university *is* dead and gone. Much of the extensive debate is in the nature of a postmortem. It is much more important to consider creating an enterprise that would embrace some positive attributes of academic thinking and harness them to the essential needs of humankind. Thinking humanity has the ability to tackle issues that were not entertained by the universities even before the decades of ruin. Our problems are global and require a global Real University to attack them.

THE REAL UNIVERSITY

We need not be grand in defining a mission statement. Let us describe as simply as possible the purpose of a Real University. The Real University will further the fundamental needs of the world's people: peace, equity, relief of poverty, creation of health, and care of the environment by means of knowledge and wisdom. The advantage of this definition is that it makes disciplines such as science, economics, and so on subservient to humanity's needs. They will be the tools of wisdom together with languages, arts, history, and philosophy.

The detailed structure, function, and role of the Real University will be described in a sister book to this text. But it will not be based on bricks, mortar, and gowns. The most familiar and far-reaching technological development of recent decades is the Internet. Like other discoveries, it can be used for good or evil. It is the facilitator and haven for pornography, pedophilia, racial vilification, fraud, illegal organizations, and much more, and for young people its video games can condition them to be killers or brainwash them into the consumer society. It offers personal empowerment in participation and enhancement of democracy, and the globalization of social forces that can be used for equity, education, environmental awareness, human rights, and for building Real Universities. The Internet and its ramifications will dwarf the industrial revolution. Its implications are already so profound that society is being transformed in its work and functions.

The existing universities have embraced the Internet with joyous alacrity, not to enhance knowledge on behalf of world peace, justice, environmental protection, or the eradication of poverty, but to compete with other financially orientated universities for distance students. These are students in faraway places who will pay for courses conducted via the Internet. It is perhaps too idealistic to wish that the motivation of these universities, together with government funding, would lead them to educate the underprivileged and the poor via the Internet.

So where should the Internet fit into the realm of higher education? It greatly enhances interaction between individual researchers and between teams of researchers. Researchers sitting in their personal offices or homes anywhere in the world can consider research concepts, data, and papers rapidly and interactively. International reports on health, justice, and the environment can be prepared by groups of academics, bureaucrats, and citizens from around the world. They can be released without prejudicial selection by publishing houses or media. The Intergovernmental Panel on Climate Change, comprising over a thousand scientists, has worked for many years to produce and evaluate information that can lead the world away from the ravages of global warming. It could be regarded as a forerunner of the working of a Real University that will provide a means for rapid interaction between groups of academics, teachers, and others on relevant topics.

It is possible to come to the conclusion that if the universities of today wish to remain degree machines linked to national economies, then the new or Real University, the connection of scholars, critics, thinkers, and public intellectuals, can and will come to exist outside these old universities. Today, with computers, the Internet, and electronic publication, interaction between thinkers and communication of thoughts and ideas is easier than it has ever been in history. Every educated person can in principle be a philosophizing Socrates in the electronic marketplace. Indeed in history many of the important advances of humankind have been outside the university system. In classical Greek philosophy, we have the example of Socrates with his philosophy of ceaseless questioning, which seems to have been practiced primarily in the marketplaces of Athens! Apart from philosophy, much of the great literature, novels, plays, and poetry of the world have been produced outside the universities. Even within science, which depends heavily upon the expensive infrastructure and laboratories of a university, major advances have occurred from thinkers working mainly in the community—Einstein, Mendel, Darwin, and Lovelock, to name but a few.

But analysis of scientific results will be important in the Real University. Consensus statements on major scientific issues will be produced. The statements will come from independent scientists. They will have ability, and their public standing and integrity will be demonstrated by declarations of their financial and scientific interests. They will have an important role in forming an international code of conduct for scientists—a Hippocratic oath for science.

But the role of the Real University will not only include evaluations of scientific and social material, it will offer new, cross-cultural, and visionary perspectives on society. The construction of an alternative society requires primarily a definition of the essential tasks fundamental to the well-being and continuation of that society. To choose examples, those tasks likely to emerge from any consensus might be reform of food production to protect the land, avoidance of pollution of waterways and coastal zones, development of new methods to ensure equitable distribution without the excessive transport costs and energy consumption. As a result, local products would be encouraged for local consumption. Other essential environmental tasks would be conservation of forests, revegetation to stabilize fragile soils and water catchments, conservation of natural waterways, and the development of natural alternatives to fossil fuels.

The principle of renewal should also extend to the urban environment where the world's older cities require extensive repair of water and sewage systems, renovation of housing for health, energy conservation, and social cohesion. All these are items at present creating environmental or other debts, which will become financial debts at some time in the future. The essential tasks of the alternative society will also include personal and population health, welfare and delivery of education, and the provision of infrastructure based upon sustainable policies. The oft-hidden philosophies of deep ecology will contribute. There has to be a recognition that a continuation of present economic growth is likely to lead to environmental calamity. Therefore there are compelling reasons to create a society of sufficiency, with the labor and resources of excessive consumption being transferred to environmental repair and care and to social cohesion. The insurmountable problem in attacking the defects of capitalism has resided in the unavailability of the alternative vision. The alternatives will evolve from an involved and interconnected community.

THE NEW BARBARIAN SOCIETY

The response of the conventionalists to these initial comments will be to cry, "We cannot afford it, we will be taxed to oblivion!" However, we should remind ourselves that money was created to allow services to be exchanged more simply. We can model a new financial system in the same way that governments model the widespread community effects of changes in taxation. So the essential tasks of alternative society can be modeled. Time and resources will be redeployed from consumption, which is normally taxed to pay for the essentials of life. But when we recognize that much of consumption is superfluous to society and causes debt for the future, we can recognize that subsequent balance sheets will be healthier. Baubles, beads, and some comforts may have disappeared, but the gains for society will be enormous. These thoughts will be developed further in the final chapter.

A negative response to change indicates that, in an age when we can travel to the moon, define the genome, and provide worldwide communications so intricate that it is difficult to grasp their complexity, we cannot or will not model and debate an alternative or modified world order? If we fail to initiate this task it is because we are so indoctrinated with the present imperfect system, or perhaps so resigned to it, that there seems no point in acting. We console ourselves with the thought that even if a nation wished to opt out of economic globalization, it could not do so, for it would surely self-destruct.

Time is short. Present economic society is evolving fast and free. This future cyberworld may have scant regard for the environmental and social needs of humanity. In *The New Barbarian Manifesto* by Ian Angell,[8] professor of information systems, future society will be composed of individualists working solely to satisfy their own economic need. This society of winners, based only on profit, will write its own rules and morality. "Knowledge workers," the innovators and generators of wealth, will be the fulcrum and leaders of modern global companies. New communities will be based around the economic well-being of these persons, their families, and friends. These information-rich global communities will be separated from the "information poor" populace both physically and mentally. The nation state will wither.

Governments and democracy will become irrelevant because their decisions cannot be implemented. Implementation requires money, and this necessitates taxation. Taxes will become progressively more difficult to increase. As tax receipts fall, pensions, hospitals, and social and environmental care will deteriorate. This process is already underway. Global industry avoids national attachment and taxes, corporation tax falls in all countries in the vain hope of attracting companies and their buccaneers. The highly paid knowledge worker or innovator has special deals on tax because of their indispensability to government. As the tax base contracts to more modest incomes, it is eroded further by the operation of e-commerce and the "black," or underground, economy. To raise tax, governments already resort to encouraging gambling and ignoring addictions such as tobacco consumption. Ultimately, the tax base will come only from food, energy, and water, the essentials of life. In this system environmental debt will continue to grow rapidly for there will be no financial ability to remedy pollution, greenhouse, salinity, or deforestation. Even our embryonic world governance systems will collapse because of unpaid dues. If Angell's vision is fulfilled the only counterbalance to this frightening scenario will perhaps reside in cyber communities of thinkers and scholars linked together from the "universities of ruins" throughout the world. This will be one format for a Real International University.

The new barbarian society will be based on information technology, skilled entrepreneurs and individualists who will form groups "based on trust, with a shared sense of enlightened self-interest, and a shared but different view of the world."[9] This new society will discard the corruption, religious bigotry, violence, and warfare of the nation states and will escape from government taxation and liberalism. Telecommunication technology will facilitate a virtual university, comprising the top universities of the United States and the world to deliver the new barbarian education. This vision is even more dangerous than the present creed of individualism because it recommends a new society with no acknowledgement of the environmental dangers facing us. It is difficult to see how these can be solved without governance and sacrifice.

History tells us that a new society, whether "barbarian" or otherwise, will revert to the self-interest and anarchy of *Animal Farm* unless humans fundamentally change their values and way of thinking. This must be the prime task for the thinkers in the Real University.

How then will the Real University fit into the spectrum of universities in the world today? Today's university, the University of Excellence, as it calls itself, will be orientated totally to producing jobs for industry, and these will be scientific, technological, managerial, and economic. It will, in effect, be a technical and vocational college without a formative role for the community. It probably is already. These universities will be heavily influenced, directed, and financed by national governments and industry, particularly multinational corporations. They will have completely abandoned a broad liberal education and what they consider to be the noneconomic subjects.

The Real University will address the fundamental problem of environmental education. At a time when environmental education is universal in early education, there is pitiful and diminishing support for environmental studies in the economically orientated universities of excellence. However environmental awareness and action in universities is unlikely to improve without fundamental change in mindsets. The Real University will have this important task, often supranational, of caring for humanity's real needs by working for true sustainability. Most importantly, it will be the potential training ground for a new generation of ecoelites who will attempt to preserve remnants of our civilization when the great collapse comes. They will be the new priesthood of the new dark age.

NOTES

1. Richard Dawkins, *The Selfish Gene* (Oxford University Press, Oxford, 1976).

2. Jane Jacobs, *Dark Age Ahead* (Random House, New York, 2004).

3. David. W. Orr, "Slow Knowledge," *Conservation Biology,* vol. 10, no. 3, June 1996, pp. 699–702.

4. N. Maxwell, *From Knowledge to Wisdom: A Revolution in the Aims and Methods of Science* (Basil Blackwell, Oxford, 1984).

5. 1 Kings 3:3–28 (King James).

6. Orr, "Slow Knowledge," from note 3, pp. 700–701.

7. Bill Readings, *The University in Ruins* (Harvard University Press, Cambridge, MA, 1996).

8. Ian Angell, *The New Barbarian Manifesto* (Kogan Page Limited, London, 2000).

9. Ibid.

— 10 —

Can Democracy Be Reformed?

We now face the prospect of a kind of global civil war between those who refuse to consider the consequences of civilization's relentless advance and those who refuse to be silent partners in the destruction. More and more people of conscience are joining the effort to resist, but the time has come to make this struggle the central organizing principle of world civilizations.

—Al Gore[1]

UNMASKING THE GLOBAL "CON"

Hundreds of scientists writing in *Millennium Assessment* and other scientific reports pronounce that humanity is in peril from environmental damage. If liberal democracy is to survive it will need to offer leadership, resolve, and sacrifice to address the problem. To date there is not a shred of evidence that these will be provided nor could they be delivered by those at the right hand of American power. Some liberal democracies that recognize that global warming is a dire problem are trying but nevertheless failing to have an impact on greenhouse emissions. To arrest climate change, greenhouse reductions of 60 to 80 percent are required during the next few decades. By contrast the Kyoto Protocol prescribes reductions of only a few percent. The magnitude of the problem seems overwhelming, and indeed it is. So much so, it is still denied by many because it cannot be resolved without cataclysmic changes to society. Refuge from necessary change is being sought in technological advances that will allow fossil fuels to be used with impunity, but this ignores the kernel of the issue. If all humanity had the ecological footprint of the average citizen of Australia or the United States, at least another three planets would be needed to support the present population of the world.[2] The ecological services of the world cannot be saved under a regime of attrition by growth economies that each year use more land, water, forests, natural resources, and habitat. Technological advances cannot retrieve dead ecological services.

The measures required have been discussed and documented for several decades. None of them are revolutionary new ideas. We will discuss the main themes of a number of important issues such as the limits to growth, the separation of corporatism and governance, the control of the issue of credit (i.e., financial reform), legal reform, and the reclaiming of the commons. Each of these issues has been discussed in great depth in the literature, and a multitude of reform movements have been spawned. Unfortunately, given the multitude of these problems and the limited resources and vision of the reformers, each of the issues tends to be treated in isolation. From an ecological perspective, which is a vision seeking wholeness and integration, this is a mistake. These areas of reform are closely interrelated and must be tackled as a coherent whole to bring about change. Banking and financial reform is, for example, closely related to the issue of control and limitation of corporate power, because finance capital is the engine of corporate expansion. The issue of reclaiming the commons and protecting the natural environment from corporate plunder is also intimately connected to the issue of the regulation of corporate power. In turn this is a legal question, and in turn legal structures are highly influenced by political and economic factors. Finally, the issue of whether there are ecological limits to growth underlies all these issues. Only if an ecologically sustainable solution can be given to this totality of problems can we see the beginnings of a hope for reform of liberal democracy. And even then, there still remains a host of cultural and intellectual problems that will need to be solved. The prospects for reform are daunting, but let us now explore what in principle is needed.

THE LIMITS TO GROWTH

Our loving marriage to economic growth has to be dissolved. The dollar value of all goods and services made in an economy in one year is expressed as the gross domestic product (GDP). It is a flawed measurement in that it does not measure the true economic and social advance of a society,[3] but it is relevant to our discussion here for most of the activities it measures consume energy. Each country aims for economic growth, for every economy needs this for its success in maintaining employment and for the perceived ever-expanding needs of its populace. Politicians salivate about economic growth, it is their testosterone boost. Most would be satisfied with 3 percent per annum and recognize that this means that the size of the economy is 3 percent greater than the previous year. On this basis the size of the economy doubles every 23 years. In 43 years it has quadrupled. Now in 23 years let us suppose that energy needs will also double in order to run this economy. Therefore if greenhouse emissions are to remain at today's level, then approximately half the energy requirements in 23 years' time will have to be alternative energy. The burgeoning energy requirements of the developing countries have not yet been included in these considerations. To date, these countries have been reluctant to consider greenhouse reductions saying that they have a right to develop without hindrance, and in any case the developed countries are responsible for most of the present burden of carbon dioxide in the atmosphere. It is not difficult to calculate therefore that there is no future for civilization in the present cultural maladaptation to the growth economy. Sustainable economic growth is an oxymoron. These arguments about doubling time apply to all other environmental calculations. Other forms of pollution that arise from the consumer society will also increase proportionally to growth, the human and animal wastes, mercury, the persistent organic pollutants, and so on. And even if some of these are ameliorated, others will arise from the activities of the burgeoning population. Science tells us that we have already exceeded the capacity of the earth to detoxify these.

In advocating a no-growth economy it has been shown in many studies that beyond the basic needs of health, nutrition, shelter, and cultural activity, which can be provided with much less income than Westerners presently enjoy, there is little correlation between wealth and happiness or well-being. A no-growth economy[4] would supply the essentials for life and happiness. Human and economic activity fuelling the consumer market would be severely curtailed and the resources redeployed to truly sustainable enterprises, basic care and repair of the environment, conservation of energy, and the manufacture of items and systems that support these needs. The standard of living as measured at present (again by flawed criteria) will fall, but there may be no alternative. The fundamental question is how can a transition be made under a liberal democracy that has consumerism and a free market as its lifeblood?

SIAMESE TWINS: THE SEPARATION
OF CORPORATISM AND GOVERNANCE

One of the founding fathers of the American Constitution, Thomas Jefferson, coined the phrase "a wall of separation between church and state" in a letter to the Danbury Baptists.[5] Whatever Jefferson exactly meant by that phrase—and there is controversy on the matter—he is generally accepted to have believed that, at least, the state should have no control over the religious beliefs of citizens. The U.S. Constitution, although not using the phrase "separation of church and state" does say, "No religious Test shall ever be required as a Qualification to any Office or Public Trust under the United States."[6] As well as "Congress shall make no law respecting an establishment of religion, or prohibiting the free access thereof."[7]

Religious writers in the United States have debated and argued long and hard about whether or not these passages exclude the establishment of a state religion or make school prayer unconstitutional. Nevertheless it is clear that there is an intention to separate the spheres of church and state, perhaps not to establish a secular society as some U.S. Christians think, but to preserve religious pluralism, diversity, and freedom. The United States after all was originally settled by the Pilgrims who left Europe, especially Britain and Ireland, so that they could practice their religious beliefs without the interference of mainstream Protestant and Catholic churches.

This separation between church and state in the United States has become quite blurred during the Bush regime, with disastrous results. As we have written, within the Republican camp there is a strong vein of Christian fundamentalism that holds that it is futile to attempt to conserve the environment because such actions merely put off judgment day. God will presumably return when the last tree has fallen. However, other fundamentalists see this position as sinful, being a failure to act as righteous caretakers of a world that God saw as good when he created it. These conflicts illustrate the importance of a secularism that bases the maintenance of civilization upon science, humanity, and wisdom.

In Australia, the Commonwealth Constitution, section 116, also states that the federal government is not to legislate in respect of religion: "The Commonwealth shall not make any law for establishing any religion, or for imposing any religious observance, or for prohibiting the free exercise of any religion, and no religious test shall be required as a qualification for any office or public trust under the Commonwealth."[8]

The situation in Europe however was essentially different from the U.S. experience. Whereas in the United States migrants sought religious freedom from the state, in Europe it was the reverse; there was a long historical process of evolution where the state sought to separate itself from religion. Under feudalism there was no separation of the church and state. The idea of the divine right of kings had it that the king was divinely appointed by God. This notion came under steady attack by various philosophers, most importantly John Locke (1632–1704) in his *Two Treatises of Government*. However the separation of the state from the church was a gradual process corresponding with the relative decline of the church's power and the emergence of capitalist society. Capitalism required such a separation for its full functioning because religious doctrines such as the ban on usury have a nasty tendency to slow the growth of capital accumulation. Secularism arose as a product of industrial capitalism and the rise of Enlightenment scientific rationality.

This is not a treatise on history. Seeking an exact historical account of how secular society developed in Europe is far beyond the scope of this book. Nevertheless the analogy is what is important here. Just as the state separated itself from religion, so must the modern state separate itself from the new religion of corporatism, for the functioning of corporatism is an anathema to truly sustainable development.

We propose that a similar separation between the private economy (i.e., big business) and the state is needed if the problems of the environmental crisis discussed in this book are to be addressed. Governments must be forced to act for political rather than economic reasons. In reality though, liberal democratic governments reflect the will of big business, and it is the corporate and financial elites who call the shots. Governments that fail to follow the agenda of corporatism are soon undermined by psychological political warfare conducted by the media. Then of course there are the threats of financial damage by financial institutions controlled by financial interests and covert or overt military action by the U.S. military against any government that *threatens* the status quo of global capitalism. William Blum in *Rogue State,*[9] has filled an entire book with well documented international examples of this. One example: Salvador Allende of Chile (1964–1973) was a popular democratically elected Marxist. The United States attempted to sabotage Allende's elections in 1964 and 1970. In 1973, having failed to destroy the government, the United States supported General Pinochet, who overthrew the Allende government. Chile was closed to the outside world for a week where over 3,000 executions and thousands of "disappearances" occurred and tens of thousands of people were tortured. Some female prisoners were raped by specially trained dogs. The then secretary of state, Henry Kissinger, said to Pinochet that "in the United States, as you know, we are sympathetic with what you are trying to do here... We wish your government well."[10]

As we have said, it is not too difficult to see how this present regime of global capitalism and liberal democracy will end: It will end through ecological necessity. Nature will take humanity by the throat and confront it with the biospherical damage that it has done. It is most unlikely in our opinion that some form of spontaneous, unorganized democratic groundswell will awaken the masses to their fates before it is too late. Rather any such resistance to the system must come from an organized vanguard, unafraid to ultimately rule in the name of the common good. These new philosopher kings feature what we call the "authoritarian alternative" discussed earlier.

It will be a core principle of any new system of government that there is a strict separation between politics and the independent financial sector. Of course the economic sphere will always be an important part of any society and will still have a political dimension. But what is required is that political decision making should be dominant over economic concerns—in other words, practical economic rationalism should be controlled by law. Along with this, it is important for governments to unite to limit by law the size and power of corporations. This will also inhibit the end point of their growth—the monopoly. Such a measure will therefore preserve competition between smaller corporations that have less power or finance to cause global environmental mayhem. The competitive analogy is seen in the salary caps for football and other team sports that prevent one or two teams from dominating the league.

The economy must be deflated in its importance in our lives and that involves—as many environmental writers have stressed—obtaining as much self-sufficiency in food and the production of the necessities of life at the local level. But more importantly it requires governments gaining control over the banking and financial sectors of society, which over the last 100 years or so have been increasingly privatized.

CAST THE MONEY LENDERS FROM THE TEMPLE

An important part of this separation between politics and the financial world is the separation of government and corporatism. Governments must cease to depend upon the corporate sector for their existence. This is nowhere more clearly seen than in the financial sphere, where the money supply of nations such as the United States and indeed most liberal democracies, lies completely in private hands. The Federal Reserve is not part of the U.S. federal government; it is a private banking institution.

Although it is not well-known by the average person, the nation's credit creation and money supply lie in the hands of private banks. Only a small quantity of society's money exists in the form of notes; most of the credit of a society nowadays is in the form of computer entries. Banks do not merely keep your money in a safe when you deposit it: rather, based on the idea that not everybody will withdraw their money at once, they are able to extend credit many times that of their actual liquid holdings (known technically as the capital adequacy ratio). Banks are then able to create credit. This black magic has given banking families enormous power and influence that has seldom been used for the common good. With globalized banking, a few financial wizards such as international financier George Soros, chairman of Soros Fund Management, are able to influence the fate of nations. This power must be taken from them, and it must be part of any sane alternative political system that the creation of credit remains the sole domain of government. A nationalization of the banking and financial sectors is as important as having a nationalized military force rather than local private armies—if not more so.

The existing damage to the world detailed in the earlier chapters cannot be remedied or even mitigated without a vast redeployment of resources from consumption to repair. Private capital will not finance this—there will be no financial return. Presently government spending of money raised from taxes is not scratching the surface of the problem. Such financing will have to be by government reclaiming a significant proportion of the creation of money. A surviving world will have massive redeployment of labor and resources from purposeless economic growth to the provision of alternative energy, water, and the conservation of land and biodiversity.

Any government, whether democratic or not, concerned with seriously tackling the environmental crisis needs to wrestle control of the nation's money supply from the private financiers. Economists in the various economic reform movements, such as COMER (Committee on Monetary and Economic Reform) have argued that not only is our economy ecologically unsustainable, but our banking system is economically unsustainable. At present, governments borrow from the private banking sector, not at the actual cost, but at many times the cost and with compound interest when none of this need occur. The banks of today are essentially printing money to create a financial system founded on debt. An individual borrowing money from a bank is not borrowing money from a depositor's account. The money borrowed is new money created by the bank. When the borrower spends this money it is released into the community. This supply of money to the community and the resulting debt forms the basis of the financial system. Money circulating in the community is created by the private banks. To quote one example, from *The Grip of Death*[11] by Michael Rowbotham, the figures for the UK in 1997 were as follows: the total money in existence was 680 billion pounds. Of this, 655 billion was created by the private banks and 25 billion by the government in the form of notes and coins. The most charitable interpretation is that this system is a giant scam whereby economies are run by the elites who pay themselves millions each year for their own benefit and that of capitalism, which depends upon debt to perpetuate consumerism. It has lead to the joke by those who understand the system: "What's the best way to rob a bank? Become the CEO." In a world requiring massive finance to repair the environment, how will this system respond? It will not. In fact many private banks in liberal democracies have a record of financing environmental destruction in the name of profit. In liberal democracies environmental repair remains a minor budget item funded, like all government expenditures, from money created by the private banks.

When this system of creating money is explained to the average citizen, there is incredulity for it is difficult to believe. When any government minister is asked about the system and its reform, there is the same shuffling silence that accompanies the question, "Can we allow economic growth to continue to infinity?" These two issues are the twin towers of complicity between the liberal democracies and the financial system that will have to be replaced if there is to be ecological survival. Perhaps the issue of the slavery of usury helped decide the targets on 9/11.

In conclusion, nations need to return to the principle of borrowing from their own central or public banks, interest free, which would be far less inflationary than borrowing from private banks. A reregulation and control of the money supply must occur. If such strategies could be implemented, then it would be possible to break down the present stranglehold that corporatism has on government.

LEGAL REFORMS: FAITH IN THE LAW?

This book has touched on the role that the legal institution has played as part of the social institution of liberal democracy in failing to adequately address the challenges of the environmental crisis and indeed in furthering environmental degradation and destruction.

One area where the law is interacting with the environmental crisis in a very direct way is in the area of climate change litigation. Climate change litigation involves plaintiffs using the mechanisms of the courts to seek redress for the alleged damages and harm caused by corporate and governmental defendants. The aim of climate change lawsuits is, in the first instance, to seek financial compensation for the environmental, human health and economic harm caused by the defendant's greenhouse gas emissions, but more importantly, to gain media coverage for the actions in an attempt to produce legal reform and a change in the behavior of governments and corporate entities.

Thus in the case of *Commonwealth of Massachusetts, et al v EPA* (2007),[12] the plaintiffs, including various U.S. states and advocacy groups, sought to have the U.S. Environmental Protection Agency (EPA) regulate greenhouse gas emissions for new motor vehicles under the Clean Air Act (1990). In July 2005, the petition was dismissed by a 2–1 decision of the court. However, in March 2006, the plaintiffs lodged an appeal with the U.S. Supreme Court. On June 26, 2006, the U.S. Supreme Court decided to hear the case and on April 2 2007 decided in favor of the plaintiffs. Also in September 2006 the State of California took action against six of America's big auto manufacturers, seeking damages for the harm that greenhouse gas emissions from their products have caused to the State of California. The case of *Massachusetts v EPA* alone constitutes a major advance in environmental law reform. Measured though in the context of the full significance of the environmental crisis, these gains are small and very hard won. For reasons that we will now detail, we do not believe that the legal institution will be at the vanguard of resolving the environmental crisis, although it does have a small contribution to make.

The legal systems of the United States and Australia are based upon English common law (the law made by judges in court decisions) and concepts. The basis of such law is that of liberalism—the ultimate worth and value of individuals. The individuals or persons involved may also be corporations—the idea of corporate personality and agency being introduced into English common law less than 200 years ago. However from the time of the Norman Conquest (1066), English law has been individual focused. The other major characteristic of the English-based systems is the supremacy given to private property rights. Offences against private property are often treated more severely than offences against individuals. As we have already seen in this book, these concepts also form the conceptual basis of capitalism. For this reason alone we doubt whether any sort of legal challenge could achieve substantial changes to the present social system. Legal challenges may produce some useful stopgap measures, but this is not enough. Outside of constitutional interpretations, governments can always make new laws to trump any court decision they do not like. Finally, legal challenge through the courts is expensive and very slow, which is far from ideal. Nevertheless in a desperate situation it is worthwhile exercising every option, and limited reforms and changes are better than nothing at all.

Environmental protection legislation in English-law based countries such as the United States and Australia has been grafted onto the individualistic private property foundation of the liberal democratic legal systems. Thus environmental protection is always balanced against competing economic interests, as we have already argued.

An ecologically sustainable legal system must give ultimate priority to the preservation of the life support systems of the earth. This value must trump the values of economic interest and personal liberty. Otherwise a tragedy of the commons situation will arise. We recall that the classical tragedy of the commons[13] is that individual economic agents operating only with principles of economic utility maximization will all pursue their exploitation of economic resource to produce the highest return, until that resource is exhausted (i.e., exterminated). The pursuit of individual self-interest results in collective environmental destruction, which ultimately threatens the life of those *individuals* and the entire economic system itself.

Therefore the *supreme* legal principle, which must be enshrined in the constitutions of all nations, must be the principle of ecological sustainability and environmental protection. Roughly drafted such a principle would assert: "This nation has an overriding legal duty to protect the environment and ensure that social, political, and all economic systems and activities that impact substantially upon the environment by any agents, persons, or entities whatsoever are ecologically sustainable." By the expression 'ecologically sustainable' we mean 'X' and in the assertion will be placed a concise drafting of the principles of sustainability. Further to that, each person and corporation has a duty of environmental protection.

The rigid enforcement of such legal principles would alone put a halt to much of the destructive development projects of modern capitalism. We have reasoned that it is inconceivable to suppose that any sort of constitutional change could be first made to produce an ecologically sustainable society. Some may therefore hope that international law could be used as a mechanism for bringing environmentally rogue nations and corporations into line. Some may hope for the creation of an international environmental court with powers greater than the international criminal court. In such a court, aggrieved parties such as the small island states, harmed by rising sea levels due to human-caused global warming, may take action against those nations that have not signed the Kyoto Protocol and its successors or who have failed to have national programs to reduce greenhouse emissions. But once again, this is faith without reason. Without a world government, it is the nation state that is the primary unit of law and international law is completely dependent upon the commitment of nation states to uphold this law. The United States and Israel, for example, have opposed the formation of an international criminal court because key political leaders could be prosecuted. The United States is no longer subject to the jurisdiction of the International Court of Justice. An international environment court "with teeth" would be rigorously opposed by the present neoconservative regime in the United States and elsewhere. It is difficult to see how judgments could be enforced against the United States and powerful corporations. At present some defamation actions against U.S. entities won in U.S. jurisdictions are not enforced in the United States because they offend U.S. constitutional protections of free speech. The difficulties in enforcing environmental actions would be even greater.

In conclusion, there needs to be major reform to the legal systems within nations and to international law if environmental damage is to be arrested. We are highly skeptical of the ability of the legal system to lead the way. Only when the larger political battle has been won or when the ecological crisis is visible to all will legal reform follow. The law is intrinsically a slow-moving, conservative beast, constructed for personal and property protection, and we cannot expect much assistance from that source. Nevertheless that is not a reason for defeatism and as environmental and human rights lawyers contrive to address these problems in the courts, we wish them well. But we are skeptical of the long-term success of these endeavors unaided by political action, and this is why our focus has been upon political and ideological change. This leads us to the question of human capacity to share the common good instead of acquiring it.

RECLAIMING THE COMMONS

"The commons" has been a recurrent theme in earlier chapters, and we return to the concept now to explain that the reforms described above are essentially those required to prevent further theft of the commons and to reclaim that already lost. In chapter 4, the retention of ecological services has been stressed as the essential factor in preserving human civilization. Ecology is part of the commons. "Eco" comes from the Greek word *oikos,* meaning house or living space for animals, plants, and humans. It has been our thesis that democratic society will self-destruct because it has placed inadequate restriction on destruction of the commons of the atmosphere, soil, and water, the natural commons. They have been enclosed by privatization or polluted to increase corporate profit by externalizing the cost of remediation. We used the easily understood example of the felling of mature forests, which causes soil erosion, pollution of waterways, and loss of water catchments, all of which are externalized to make profits larger. Ultimately the community pays.

The environmental commons is the shared wealth of humanity, but we should recognize that the commons has a wider connotation and includes commons already divested to private ownership. It includes government and academic research discoveries financed by community taxes and then handed over to corporates. It includes the control of agricultural seed lines and patents for communal foods. It includes control of the airwaves handed out as political patronage and control of the Internet discovered by government and now taken i   te ownership.[14] Included in this litany of lost resources are the ban   money supply. We have discussed the mechanisms by wh   red these community resources with the enthusia   democracies. The actions have been rationalize   ation efficiency, smaller government, and r   this charge with a new assault by the   rces" that promises to strik   omic use. This is a recurring   essive "balance" the commor   rative in all liberal de-mocrac   an in the United States. Lib   ism with its excesses of capi   rshipping at this altar even fai   row confines of their ideol-ogy.   ontrols in 50 of the U.S. states and show   in economic weakness. Envi-ronmental prote   , more equality of wealth, better public health, and fair   se it requires bigger government and results in less freedom an   the corporates, it is not acceptable on ideological grounds. Conseque   is a common occurrence for those promoting regulation, often enviror   alists, to be labeled as "activists" or "socialists" and to be accused of being responsible for the loss of jobs. A wedge is driven between them and their natural allies—the workers who are concerned about their quality of life.

Retrieval of the commons requires the monetary and legal reforms outlined in the present chapter. There is no evidence that self-regulation by corporates is effective. If there is to be any raison d'etre for retaining liberal democracies, they have to rigidly control corporate activities nationally and internationally and control the money supplied to finance environmental repair. We come therefore to the fundamental question as to whether human nature has the will to do this under the present system of governance.

HUMAN CAPACITY AND METAMORPHOSIS

A study of human history, intelligence, psychology, and capacity to change is perhaps the most important component in these deliberations.

We can learn from the experience of earlier civilizations, some of which have self-destructed while others flourished. Jared Diamond in *Collapse: How Societies Choose to Fail or Survive* indicates the five interactive factors that brought collapse to specific civilizations.[17] They are environmental damage, climate change, changes in friendly trading partners, and the society's political, economic, and social responses to these shifts. The Polynesian society on Easter Island died because they destroyed their environment. The Norse colonies on Greenland succumbed to climate change, enemies, and problems with trade. The demise of the Mayan, Mesopotamian, and Roman civilizations, while complex, had significant environmental causes. Diamond lists other societies in Japan, parts of Europe, and the New Guinea highlands that have survived for prolonged periods of time because they recognized their problems and instituted societal or environmental change. In history each of these civilizations was localized geographically and could not benefit from the adverse experiences of others. Today we can examine the failures and learn from them. Or can we?

In this text we have listed the impediments to recognition within the liberal democracies, the inertia and self-interest in preventing political change, the self-interest of the so-called free press, and the corporate and financial interests. And to these impediments we must add the lack of understanding by the ordinary robotic worker and mechanical consumer that he or she has now become. It will require a fundamental change in society for the citizen to be able to understand the present political system, let alone the complexities of our dependence on ecological services. We doubt if any transformation of the masses is possible, at least to the extent needed for a radical democratic transformation of the present system. For example, most people have difficulty understanding the nature of the monetary system of capitalism at the basic level described here. It is difficult even for those with slightly higher IQs to grasp the diabolical logic of credit creation. Yet without such a grasp, reform of the present system is impossible. Without leadership with a will and power to act the crisis is certainly insoluble.

As scientific realists, we must look elsewhere if we are to find a political answer adequate to the challenge of the environmental crisis. Democracy, like communism, is a nice idea, and it is a pity that neither works. If there was a way of saving democracy then we should save it, but it is unlikely that there is any such way because the ordinary person or "mass man" is not made of the right heroic stuff necessary to meet the challenge of our age.

Should we have faith in the capacity of humanity to reform its thought processes, and could generational change bring reform? There is one global civilization today and the crisis is global. How can the world's population change its attitudes? The brain of each person, despite its instinctual behavior discussed in chapter 6, has a capacity to evolve to encompass new thinking and skills. This is seen in the brain of the musician that expands and develops its musical area. The youngest generation of humanity has changed its mode of thinking and communication by involvement in the electronic revolution.[18] The youngest generation, the "digital natives," are "inventing new online ways of making each activity happen based on the new technologies available to them."[19] The activities that have changed are cited as communication, buying and selling, creativity (games), meeting, collecting, learning, socializing, and many more. Perhaps this is the new definition of freedom allowed within a society of proles, in effect an electronic cage enclosing the brain and the computer in one circuit-firing mass. For despite the liberating reach of the web, this is a generation without protest or activism that has known no other type of society. A generation robotically tied to an economic system, career, university fees, and two-salary homes to assuage the consumer brain in the status anxiety of Western culture. A chattering generation that cannot envisage the flip side of the social universe or a brain that might traverse a black hole to emerge into a different value system.

This is not to say that the Internet cannot be the fulcrum of change. It has proven indispensable for the organization of global protest over the actions of corporatism and at the meetings of the World Trade Organization and other financial summits. A common thread of concern for human rights, poverty, and the environment has coalesced interest groups into unified demonstration. But they have been contained as anarchists by governments and the press and are irrelevant to the majority who wish to remain uninterrupted in their culture of consumption. So far they are impotent in the containment of the excesses of capitalism.

Some have enthused that this organizational weapon provides an opportunity to reform democracy through informed participation: the World Social Forum and other grassroots meetings can evolve.[20] We have argued that it is too late to rely upon an evolution of true democracy through bottom-up reform. The performance of humanity to date suggests that we will move toward crisis, followed by disorder and authoritarian rule.

Preparation for the future is the vital role of education. Not the form of education peddled by the higher education establishments of Western civilizations, for these have become the tools of our maladaptive economic culture, but an education for living within the confines of the earth's supporting systems. A new, unsullied system will have to be created that conveys knowledge on sustainability—correct, uncensored, unedited, and scientifically correct knowledge. A Real University education obtained willingly and freely by those wishing a new Enlightenment will provide the technocratic leaders of the future. Though this reform may be too slow to assuage the growing environmental problems, we must try.

An artist will say that painting a picture is a journey with an unknown destination until you arrive. The authors, inculcated with democracy and its culture, commenced their journey through the matrix of ecology and governance and arrived at an unexpected destination. The completed picture has changed our vision. To conclude, we ask you the reader to examine evidence that the political system under which you live has implemented any significant reform that will arrest the environmental crisis or indeed achieve any one of the goals described in this chapter. Further, do you think that liberal democracy has the capacity and resolve to do so? You will then be able to decide whether you have faith in liberal democracy to manage this crisis. The authors suspect that you will read chapter 8 once more.

NOTES

1. Al Gore, *Earth in the Balance* (Houghton Mifflin, Boston, 1992), p. 294.

2. M. Wackernagel and W. Rees, *Our Ecological Footprint: Reducing Human Impact on the Earth* (New Society Publishers, Gabriola Island, BC, Canada, 1996).

3. D. Shearman and Gary Sauer-Thompson, *Green or Gone: Health, Ecology, Plagues, Greed and Our Future* (Wakefield Press, Adelaide, Australia, 1997).

4. C. Hamilton, "The Post-Growth Society," in C. Hamilton (ed.), *Growth Fetish* (Allen & Unwin, Sydney, 2003), pp. 205–240.

5. Thomas Jefferson, "Letter to the Danbury Baptists," 1802, at <http://www.usconstitution.net/jeffwall.html>.

6. U.S. Constitution, Article VI, Sec, 3.

7. U.S. Constitution, Bill of Rights, First Amendment.

8. In the case of *Attorney-General (Vict) (Ex.Rel. Black) v. Commonweath* (1981) 146 CLR 559, Wilson J. and Stephen J. both held that section 116 did not guarantee within Australia the strict separation of church and state. Mere non-establishment does not logically imply a strict separation of church and state. See M. Wallace, "Is there a Separation of Church and State in Australia and New Zealand?" *Australian Humanist,* No. 77, Autumn 2005.

9. William Blum, *Rogue State* (Zed Books, London, 2002).

10. Ibid., p. 143.

11. Michael Rowbothom, *The Grip of Death. A Study of Modern Money, Debt Slavery and Destructive Economics* (John Carpenter Publishing, Charlbury, England, 1998).

12. See J. W. Smith and D. Shearman, *Climate Change Litigation: Analysing the Law. Scientific Evidence and Impacts on the Environment, Health and Property* (Presidian Legal Publications, Adelaide, Australia, 2006); *Commonwealth of Massachusetts et al v EPA,* 1275 S Ct. 1438, 549 US (2007).

13. G. Hardin, "The Tragedy of the Unmanaged Commons," *Trends in Ecology and Evolution,* vol. 9, 1994, p. 199.

14. David Bollier, *Silent Theft. The Private Plunder of Our Common Wealth* (Routledge, New York, 2002).

15. A. Saad-Filho and D. Johnson, *Neoliberalism. A Critical Reader* (Pluto Press, London, 2004).

16. Paul H. Templet, *Defending the Public Domain: Pollution, Subsidies and Poverty,* PERI Working Paper No. DPE-01-03 (University of Massachusetts, Political Economy Research Institute, 2001).

17. Jared Diamond, *Collapse: How Societies Choose to Fail or Survive* (Allen Lane, Camberwell, Australia, 2005).

18. Mark Prensky, "The Emerging Online Life of the Digital Native," 2004, at <http://www.marcprensky.com/writing/default.asp>.

19. Ibid.

20. N. Klein, *Fences and Windows: Dispatches from the Front Lines of the Globalisation Debate* (Flamingo, London, 2002).

Bibliography

Abbey, E., *The Monkey Wrench Gang* (Harper Perennial Modern Classics, New York, 2000).

Alverson, D.L., and Dunlop, K., *Status of World Marine Fish Stocks* (University of Washington School of Fisheries, University of Washington, 1998).

Angell, M., *The Truth About Drug Companies: How They Deceive Us and What To Do About It* (Random House, New York, 2004).

Appleyard, B., *Understanding the Present: Science and the Soul of Modern Man* (Picador/Pan Books, London, 1992).

Bakan, J., *The Corporation: The Pathological Pursuit of Power and Profit* (Constable and Robertson Ltd, London, 2004).

Beder, S.D., "Corporate Highjacking of the Greenhouse Debate," *The Ecologist*, vol. 29, 1999, pp. 119–122.

Belloc, H., *The Servile State* (T.N. Foulis, London, 1912).

Bertel, R., Dyer, K., and Gray, B., "Is Christianity Green? A Critique of Some Accepted Views on the Relationship Between Christianity and Environmentalism: A Discussion Paper," (Mawson Graduate Centre for Environmental Studies, University of Adelaide, Australia, 1995).

Blum, W., *Killing Hope* (Zed Books, London, 2003).

Blum, W., *Rogue State* (Zed Books, London, 2002).

Bollier, D., *Silent Theft: The Private Plunder of Our Common Wealth* (Routledge, New York, 2002).

Boyden, S., *The Biology of Civilisation: Understanding Human Culture as a Force in Nature* (University of New South Wales Press, Sydney, 2004).

Britton, S., "The Economic Contradictions of Democracy," *British Journal of Political Science*, vol. 5, 1975, pp. 129–150.

Brown, H.O.J., "Cultural Revolutions," *Chronicles,* June 2001, pp. 6–7.

Bryden, H.L., et al., "Slowing of the Atlantic Meridional Overturning Circulation at 25 N," *Nature*, vol. 438, 2005, pp. 655–657.

Burkhart, R.E., and Lewis-Berk, M.S., "Comparative Democracy: The Economic Development Thesis," *American Political Science Review,* vol. 88, 1994, pp. 903–910.

Burnham, J., *Suicide of the West: An Essay on the Meaning and Destiny of Liberalism* (Jonathan Cape, London, 1965).

Campbell, C.J., *The Essence of Oil and Gas Depletion* (Multi-Science Publishing Co Ltd, Essex, England, 2004).

Campbell, C.J., *Oil Crisis* (Multi-Science Publishing Co Ltd, Essex, England, 2005).

Conner, S., "Scientists Warm to Hurricane Theory," *The Independent Weekly*, December 11–17, 2005, p. 10.

Dahl, R., *Modern Political Analysis* (Prentice Hall, Englewood Cliffs, NJ, 1991).

Daly, H.F., and Cobb Jr., J.B., *For the Common Good*, 2nd edition (Beacon Press, Boston, 1989).

Dawkins, R., *The Selfish Gene* (Oxford University Press, Oxford, 1976).

de Botton, A., *Status Anxiety* (Pantheon Books, New York, 2004).

de la Boetie, E., *The Politics of Obedience: The Discourse of Voluntary Servitude* (Free Life Editions, New York, 1975).

Dennis, L., *The Coming American Fascism* (Harper and Brothers Publishers, New York, 1936).

Dershowitz, A., *The Case for Israel* (Wiley, New York, 2004).

Diamond, J., *Collapse: How Societies Choose to Fail or Survive* (Allen Lane, Camberwell, Victoria, Australia, 2005).

Ehrlich, P.R., and Ehrlich, A.H., *Healing the Planet: Strategies for Resolving the Environmental Crisis* (Addison-Wesley, Reading, MA, 1991).

Essex, C., and McKitrick, R., *Taken by Storm: The Troubled Science, Policy and Politics of Global Warming* (Key Porter Books Limited, Toronto, 2002).

Fahn, J.D., *A Land on Fire: The Environmental Consequences of the Southeast Asian Boom* (Westview Press, Boulder, CO, 2003).

Feyerabend, P.K., *Against Method* (SCM Press, London, 1975).

Feyerabend, P.K., *Killing Time* (University of Chicago Press, Chicago, 1995).

Gelbspan, R., *Boiling Point* (Basic Books, New York, 2004).

Georgescu-Roegen, N., *Economic Theory and Agrarian Economics* (Oxford Economic Papers, Oxford, 1950).

Glendon, M.A., *Rights Talk: The Impoverishment of Political Discourse* (The Free Press, New York, 1991).

Goodstein, D., *Out of Gas: The End of the Age of Oil* (W.W. Norton, New York, 2004).

Gore, A., *Earth in the Balance* (Houghton Mifflin, Boston, 1992).

Gottfried, P., *After Liberalism: Mass Democracy in the Managerial State* (Princeton University Press, Princeton, NJ, 1999).

Graham, G., *The Case Against the Democratic State* (Imprint Academic, Charlottesville, VA, 2002).

Gray, J., *Straw Dogs: Thoughts on Humans and Other Animals* (Granta Books, London, 2002).

Gurr, T.R., "Persistence and Change in Political Systems, 1800–1971," *American Political Science Review*, vol. 68, 1974, pp. 1482–1504.

Hamer, D.H., *The God Gene* (Doubleday, New York, 2004).

Hardin, G., "The Tragedy of the Commons," *Science*, vol. 162, 1968, pp. 1243–1248.

Heinberg, R., *The Party's Over: Oil, War and the Fate of Industrial Societies* (New Society Publishers, Gabriola Island, BC, Canada, 2003).

Hoppe, H-H., *Democracy: The God that Failed: The Economics and Politics of Monarchy, Democracy, and Natural Order* (Transaction Publishers, New Brunswick, NJ, 2001).

Jacobs, J., *Dark Age Ahead* (Random House, New York, 2004).

Kaplan, R., "Was Democracy Just a Moment?" *The Atlantic Monthly*, December 1997, pp. 55–80.

Kohr, L., *The Breakdown of Nations* (Reinhart, New York, 1957).

Kunstler, J.H., *The Long Emergency* (Atlantic Books, London, 2005).

Leggett, J., *The Empty Tank* (Random House, New York, 2005).

Lovelock, J., *The Revenge of Gaia* (Allen Lane, London, 2006).

Ludovici, A.M., *The Specious Origins of Liberalism: The Genesis of an Illusion* (Britons Publishing Company, London, 1967).

Lutton, W., and Tanton, J., *The Immigration Invasion* (The Social Contract Press, Petoskey, MI, 1994).

MacIntyre, A., *After Virtue* (Duckworth, London, 1981).

Maxwell, N., *From Knowledge to Wisdom: A Revolution in the Aims and Methods of Science* (Basil Blackwell, Oxford, 1984).

McMichael, A.J., *Human Frontiers, Environments and Disease* (Cambridge University Press, Cambridge, 2001).

Mencken, H.L., *A Mencken Chrestomathy* (Vintage Books, New York, 1982).

Monbiot, G., *The Age of Consent: A Manifesto for a New World Order* (Flamingo, London, 2003).

Monbiot, G., *Heat: How to Stop the Planet Burning* (Allen Lane, London, 2006).

Norris, P., and Inglehart, R., *Sacred and Secular: Religion and Politics Worldwide* (Cambridge University Press, Cambridge, 2004).

Olin, S.M., "Feminism and Multiculturalism: Some Tensions," *Ethics*, vol. 108, 1988, pp. 661–684.

Ophuls, W., *Ecology and the Politics of Scarcity: Prologue to a Political Theory of the Steady State* (W.H. Freeman, San Francisco, 1977).

Ophuls, W., *Requiem for Modern Politics: The Tragedy of the Enlightenment and the Challenge of the New Millennium* (Westview Press, Boulder, CO, 1997).

Ophuls, W., and Boyan Jr., A.S., *Ecology and the Politics of Scarcity Revisited: The Unravelling of the American Dream* (W.H. Freeman, New York, 1992).

Orwell, G., *Nineteen Eighty-Four* (Penguin Books, Middlesex, England, 1954).

Perelman, L.J., "Speculations on the Transition to Sustainable Energy," *Ethics*, vol. 90, April 1980, pp. 392–416.

Potts, M., and Short, R., *Ever Since Adam and Eve: The Evolution of Human Sexuality* (Cambridge University Press, Cambridge, 1999).

Readings, B., *The University in Ruins* (Harvard University Press, Cambridge, MA, 1996).

Rushton, P., *Race, Evolution and Behaviour*, 3rd edition (Charles Darwin Research Institute, Port Huron, MI, 2000).

Sharansky, N., *The Power of Freedom to Overcome Tyranny and Terror* (Public Affairs, New York, 2004).

Shearman, D., "Time and Tide Wait for No Man," *British Medical Journal*, vol. 325, 2002, pp. 1466–1468.

Shearman, D., and Sauer-Thompson, G., *Green or Gone: Health, Ecology, Plagues, Greed and Our Future* (Wakefield Press, Adelaide, Australia, 1997).

Smith, J., and Shearman, D., *Climate Change Litigation: Analysing the Law, Scientific Evidence and Impacts on the Environment, Health and Property* (Presidian Legal Publications, Adelaide, Australia, 2006).

Somit, A., and Petersen, J.A., *Darwinism, Dominance and Democracy* (Praeger, Westport, CT, 1997).

Stiglitz, J., *The Roaring Nineties* (Penguin Books, London, 2003).

Tabor, G.M., and Aguirre, A.A., "Ecosystem Health and Sentinel Species: Adding an Ecological Element to the Proverbial 'Canary in the Mineshaft,'" *EcoHealth*, vol. 1, 2004, pp. 226–228.

Thomas, C.D., et al., "Extinction Risk from Climate Change," *Nature*, vol. 427, 2004, pp. 145–148.

Vanhanen, T., *The Emergence of Democracy* (The Finnish Society of Science and Letters, Helsinki, 1984).

Wackernagel, M., and Rees, W., *Our Ecological Footprint: Reducing Human Impact on the Earth* (New Society Publishers, Gabriola Island, BC, Canada, 1996).

Weale, A., "The Impossibility of Liberal Egalitarianism," *Analysis*, vol. 40, 1980, pp. 13–19.

Wilson, E.O., *The Future of Life* (Little Brown, London, 2002).

Wolff, R.P., *In Defense of Anarchism* (Harper and Row, New York, 1970).

Zakaria, F., *The Future of Freedom: Illiberal Democracy at Home and Abroad* (W.W. Norton, New York, 2003).

Index

About the Authors

DAVID SHEARMAN is a physician and scientist who has held positions on Faculty at Edinburgh and Yale University Medical Schools and as Mortlock Professor of Medicine at Adelaide University. He is presently working on several aspects of health and climate change. As a practicing physician he has maintained a lifelong interest in the environment and its relationship to human health.

JOSEPH WAYNE SMITH is a lawyer and philosopher with a research interest in environmentalism. He is the author of *Global Meltdown* (Praeger, 1998) and *Healing in a Wounded World* (Praeger, 1997), among other books on the future of human civilization.